To Roger Asch —

May you never drill
a dry hole in life!

Kindest personal regards,
Ruth Sheldon Knowles
May, 1982

THE GREATEST GAMBLERS

The
Epic
of
American
Oil
Exploration

THE GREATEST GAMBLERS

Ruth Sheldon Knowles

Second Edition

UNIVERSITY OF OKLAHOMA PRESS

NORMAN

To all the unsuccessful explorers who have drilled
America's more than 730,000 dry holes and
whose failures have guided others to the discovery
of the abundance of oil and gas which made
America a great industrialized nation, I dedicate
this book with admiration. Their courage,
venturesomeness, faith, tenacity, and optimism
continue to inspire today's explorers who
persist in the search to find and develop the
nation's remaining great oil and gas potential.

Books by Ruth Sheldon Knowles

The Greatest Gamblers: The Epic of American Oil Exploration
(New York, 1959; Norman, 1978)
Indonesia Today: The Nation That Helps Itself (Plainview, N.Y.,
1973)
America's Oil Famine: How It Happened and When It Will End
(New York, 1975)
America's Energy Famine: Its Cause and Cure (Norman, 1980)

Preface

As i was finishing the last chapter of the first edition of this book, my five-year-old daughter Nora, watching me type, asked, "How many miles have you done on your book, Mommy?" I nearly answered: "All my life." I feel particularly fortunate that I was born in the oil patch and have spent my life learning about the oil and gas industry as a journalist, petroleum specialist, independent oilwoman, and consultant to governments and industry. As a child, before I ever found out that a wildcat was an animal, I knew it was a well that my Oklahoma independent oilman father was drilling in search of oil, hoping to pay last month's bills with the oil he was going to find tomorrow. I am grateful that God put oil in so many different places in the world, because it has given me the opportunity to follow the oil seekers and finders to deserts, jungles, mountains, oceans, and the Arctic Circle, seeing oil fields, pipelines, tankers, refineries, and research laboratories. I learned about the politics of oil working with governments, studying petroleum laws, attending Congressional hearings and Arab oil congresses.

Most of all, however, my finest source of inspiration has been the great wildcatters and scientists who developed the art of oil exploration, and they are what this book is about. Emerson says, "There is not a property in Nature but a mind is born to seek and find it." The minds who were born to seek and find oil are among the most fascinating and productive in all history. They changed the destiny of America and gave it the energy with which to build the world's greatest industrialized nation. Their imagination, courage, and ingenuity made the story of petroleum the greatest romance in industrial history.

Strangely enough, exploring for oil did not become truly important until 1901, when the Spindletop gusher in Texas made men conscious of the abundance of oil in the earth. Almost 98 per cent of all the oil found in America has been discovered since then, so that the majority of those who forged this vigorous new art were men of recent times. It was my privilege to know many of these men well, to have had them for my teachers, with participation in the industry as my classroom. Much of the material in this book was told to me by the explorers themselves.

Although this book was first published in 1959, it is even more pertinent today. The past decade has been bewildering for Americans. We were not prepared for an abrupt transition from an economy of cheap energy abundance to one of shortages, crises, high prices, confusing domestic and international politics, discreditation of the oil industry, and sharply conflicting opinions about our energy present and future. It has been like driving a car at high speed and hitting a huge mud puddle, with the windshield becoming so smeared that the road is no longer visible. A look at our energy past will do much to clarify our vision. In our journey so far, the greatest gamblers proved that the way to solve problems was through self-confidence, resourcefulness, persistence, and the capacity to turn seeming defeats into new opportunities for success. The road we have travelled still stretches ahead, and the signposts to guide us remain the same.

As I was finishing the new last chapter of this edition in order to bring the story up to date, it became clear to me that, despite our present difficulties, America's energy future is still bright with prospects for self-reliance and growth. Other energy sources will be developed, but the Petroleum Age is far from ended. Our earth scientists—although not the politicians—agree that there are great oil and gas resources remaining to be found. And the greatest gamblers are still among us to find them if they are provided economic incentive. The names are changing but the breed is not. Today's wildcatters and scientists face great odds and frustration, but so did those who pioneered the art. As E. DeGolyer, the father of petroleum geophysics, pointed out, it is "dynamic" art constantly evolving. In the final analysis, the extent of our resources can be measured only by our resourcefulness and *that* is a quality which we possess in abundance.

RUTH SHELDON KNOWLES

New York City

Contents

BOOK ONE

Seeds
of Time

If you can look into the seeds of time,
And say which grain will grow and which will not,
Speak then to me, who neither beg nor fear
Your favors nor your hate.

(Macbeth)

1 : *"He shook the boughs . . ."*

ONE SATURDAY AFTERNOON in August, 1859, a dark, bearded man in a stovepipe hat and frock coat sat on a log beside a creek in a Pennsylvania valley, listening moodily to the rhythmic sound of an iron tool pounding a hole through rock. Edwin L. Drake was depressed and discouraged. His whole life had been a chain of petty failures, and the machinery nearby seemed to be forging yet another link, instead of drilling a well in search of rock oil.

How eagerly he had snatched the opportunity to come to Titusville the year before, believing that the Connecticut banker who suggested it had opened a door into a new way of life for him. He had even invested his life's savings of $200 in stock of the new Pennsylvania Rock Oil Company, whose operations he was now directing. But the venture, he thought bitterly, was going to prove as worthless as his unwarranted and unexpected title of Colonel. He should have become suspicious when the company backers began sending letters addressed to Colonel Edwin L. Drake, explaining casually that the title would lend dignity and create confidence in the enterprise. True, he had worn a uniform before coming to Titusville—that of a railroad conductor. No one suspected that his only other battles during his forty years had been dismal struggles as a steamboat night clerk, farm laborer, hotel clerk, and dry-goods store salesman.

There was no dignity left in the title of Colonel now— witness the bag of flour which he had just accepted from the sympathetic owner of the land on which he was drilling, who knew he didn't have any money for food. Was it the title or his own efforts that had induced the bank to loan him $500

3

a few weeks before to continue drilling when his company refused to send any more money? Merchants had given him credit and even loaned him a horse.

Why should he try to carry this burden alone? he asked himself. Why not accept the fact that this was just another one of his failures?

In the beginning he had been certain that this unique, exciting plan would make him a fortune. The circumstances under which the idea originated had unfolded naturally, with sound recommendations and logical, but imaginative, business thinking.

Dr. F. B. Brewer, a Titusville physician, visiting his professor of surgery at Dartmouth College, had showed him as a curiosity a sample of oil skimmed from a spring on land in which his family owned an interest. The professor in turn showed the sample to another Dartmouth graduate, George Bissell, a New York lawyer. Bissell wondered if the oil could be used for lamp fuel in place of the standard and cheapest illuminant, whale oil. Once plentiful, the supply of whale oil was no longer adequate. Benjamin Silliman, a Yale University chemistry and geology professor, analyzed Bissell's sample and declared it could be refined into excellent kerosene.

Bissell immediately leased the Pennsylvania land on Oil Creek containing the spring and interested some New Haven bankers in backing a company that would try to develop oil in quantities. If he had not chanced to see a certain patent-medicine circular, he would probably have recommended digging shallow pits as men had been doing for thousands of years in Asia, Africa, and Europe at places where oil seeped out, using it to calk boats, treat ailments, make things stick together and burn in lamps. However, Bissell's eye was caught by the circular's picture of a wooden tower housing machinery used to drill salt wells. Samuel Kier, a Pittsburgh salt-works owner, had found a profitable use for an annoying by-product—the rock oil that came up with the brine. He advertised it as "Kier's Petroleum or Rock Oil Celebrated for its Wonderful Curative Powers. A Natural Remedy Procured from a Well in Allegheny County, Pennsylvania, 400 feet below the Earth's Surface."

If oil were found indirectly as a result of drilling for salt, Bissell wondered, why not use the same method to find oil directly? It did not occur to the Pennsylvania Rock Oil Company promoters that any special talent would be required to drill such a well or they would probably not have selected so unlikely a manager as a railroad conductor.

Nor did the newly commissioned colonel think there would be any problems. In Titusville, he hired Uncle Billy Smith, a sprightly old blacksmith, knowledgeable about salt wells and salt works, who competently organized a "rig raising," a jovial affair like an old-fashioned barn-raising. Within an hour, some thirty volunteers built the wooden derrick and shed to house the big wheel with its coiled cable, which, powered by steam, would roll up the cable suspended from a pulley high in the derrick, then release it so that the iron bit attached to the end of the cable would strike powerful blows, crushing and pounding a hole through the rocks.

Uncle Billy and his two sons kept the hole going down while Colonel Drake supervised and fished and played euchre with Titusville merchants. However, it was not long before Uncle Billy ran into trouble. The rocks through which they were drilling were water reservoirs, and more time had to be spent bailing out water than drilling. At the end of a year of laboring, the hole had not quite reached 40 feet and had cost $2,490. The New Haven promoters, unable to sell more stock, refused to send Drake another dollar despite the fact that he assured them he had finally solved the drilling problem.

It was his idea to put lengths of pipe down the hole to the point where the hole was below the flowing water formations, then, with the water effectively shut out, continue drilling the hole inside the pipe. Edwin L. Drake was no longer an imposter. He had earned a real title—that of inventor—and his ingenious idea would be adopted by all hole-drillers after him.

"Colonel, it's quitting time," Uncle Billy called cheerfully from the rig floor.

He doesn't realize how accurate that statement is, Drake thought. The hole had been going down at the rate of 3 feet a day, but now there was no more money.

"How deep are we?" Drake asked his driller.

"As near as I can reckon, sixty-nine and a half feet," Uncle Billy said. "Well, tomorrow's Sunday. We won't drill again till Monday, but I'll come back out in the morning and tidy up around here."

Wearily, Drake said good night and went back to Titusville.

Next morning Uncle Billy, puttering around the well, happened to look down the hole. He noted with irritation that water had seeped up to within 10 feet of the top. That would mean plenty of bailing next morning. As he peered closer he saw something dark on the water. He tied a rope on a water can and lowered it in the hole. It came up dripping with oil.

Uncle Billy hurried to town and slyly did not tell Drake why he wanted him to come back to the well with him.

"Look there!" Uncle Billy said. "What do you think of this?"

The mystified Drake looked down the hole. "What's that?" he asked.

"That's your fortune," Uncle Billy said triumphantly.

Here was the unspectacular beginning of an industry soon to be trademarked by its gushers. Thirty years before, a salt well on the banks of Kentucky's Cumberland river had spurted oil 30 feet into the air at the rate of 1,000 barrels a day. Oil covered the river, which, catching fire, flowed a solid sheet of flame for 40 miles. It is not strange that no one's imagination caught fire at such a dramatic sight, for there was then a seemingly inexhaustible supply of cheap whale oil for America's lamps and there were no big factories demanding fuel and lubrication for their machinery.

Drake's well oozing oil dated the birth of the oil industry simply because it was the first well that sought oil and found it, at a time when the world urgently needed an abundant new source of illumination, fuel, and lubrication.

When Drake rigged up a pump and began producing 35 barrels of oil a day, selling each barrel for $40, that was all the stimulus anybody's imagination needed.

There was a rush to punch holes along Pennsylvania creek banks that rivaled the California Gold Rush ten years earlier. Those who did not have drilling rigs "kicked down" wells by the ancient Chinese spring-pole method. A rope with drilling tool attached ran over a pulley suspended in a crude derrick. The rope was fastened to the end of a long spring pole whose other end was anchored some distance from the derrick. Two men placed their feet in a strap suspended from the rope and, by kicking, bent the pole down. Its natural spring, after release, moved the drilling tool up and down. Drilled in such a crude, laborious way, wells on Oil Creek gushed as much as several thousand barrels a day. Anyone with strong leg muscles could make his fortune, and almost everyone did except Edwin Drake.

For him, the historic Titusville well proved to be, as he had feared, another of his petty failures. He did not apply the same imagination to opportunity that he showed in solving the problem of drilling. While his associates in the Pennsylvania Rock Oil Company independently took leases on other properties and became wealthy, Drake accepted the job of Titusville justice of the peace, notarizing other men's oil leases and buying and selling other men's oil on commission. He showed even less imagination about the possible uses of the abundant new natural resource than an old farmer who visited the original well and asked for a sample, saying, "I think it would taste durn good on pancakes."

The forest of wooden derricks bled the Pennsylvania earth of such quantities of oil that within two years it sold for 10 cents a barrel, demoralizing the new industry almost at its inception. Drake, taking the few thousand dollars he had accumulated, moved to New York, once again looking for that open door to a new way of life. A few years later a Titusville friend encountered him in Wall Street, ill and penniless. The shocked friend called a public meeting of oilmen in Titusville and raised $4,200 for the man who had started them all.

The Pennsylvania legislature eventually recognized the state's debt to Drake as the creator of its great new wealth and passed an act granting him an annuity of $1,500; after his death in 1880 it was continued to his widow.

Dr. F. B. Brewer, who had unwittingly godfathered the

new industry by showing the sample to his Dartmouth professor, was no more enriched by subsequent events than Drake was. Brewer's faith in the outcome of Drake's well had been so scant that while it was drilling he traded his Pennsylvania Rock Oil shares for cigars.

The summation of Edwin Drake's career that Brewer later provided was more to the point than the Homeric platitudes engraved on an elaborate Grecian tombstone donated by a wealthy oilman twenty-one years after Drake's death.

"He shook the boughs," Brewer said simply, "for others to gather the fruit."

2 : "... your young men shall see visions"
(Joel)

JOHN D. ROCKEFELLER, a young Cleveland, Ohio, merchant who dealt in meat, grain, and produce, was intrigued by certain prospects of the exuberant new industry that had been born in Pennsylvania. The risky business of drilling had no appeal for him. What stimulated his thinking was the knowledge that a gallon of kerosene sold for twice as much as a whole barrel of crude oil.

As the drillers on Oil Creek proved the abundance of raw material, refineries mushroomed at terminal points of rivers and railroads leading from the Oil Region. Cleveland was the country's major shipping center to the East and was also on the main railroad. Four years after Drake's discovery at Titusville, the Ohio city had twenty refineries.

The nation, emerging from the Civil War, was headed for boom times. Europe was clamoring for American kerosene.

The manufacturing of oil products seemed to offer a secure field for a resourceful, cautious, yet ambitious young man. Rockefeller invested in a refinery, tested its profit-making ability, and enthusiastically dedicated himself to oil. As an eleven-year-old on his parents' New York farm he had shown such an unusual money-making capacity in routine things such as turkey raising that his elder sister, Lucy, sagely observed, "When it's raining porridge you'll find John's dish right side up." It was raining oil now, and John was prepared. His talent for acting boldly, but never imprudently, was so great that he and his associates quickly gained control of the industry, his Standard Oil Company emerging as the largest and richest manufacturing company in the world. As Rockefeller shrewdly foresaw, control would belong to whoever could obtain a throttlehold on transportation and refining. By 1880, owning 80 per cent of America's refining capacity and 90 per cent of her pipelines, Standard Oil dictated railroad policies and dominated world marketing. Competition was virtually eliminated. As for the oil finders and producers—let them mismanage, be profligate, glut the market, and fight among themselves. They had only Standard Oil to sell to, and Standard had brought "order out of chaos" —Rockefeller's euphemism for monopoly.

How Rockefeller achieved his monopoly makes astonishing reading in the light of present-day business ethics and laws. Secret rebates from railroads, extinction of one small company after another, intimidation, bribery, collusion, deception—even his apologists admit his use of all these weapons, and his critics accuse him of far worse. He was no different in his methods than any other rising tycoon of his times—just more successful. There were no restrictions then.

Rockefeller firmly believed he was working for the whole

industry's good by organizing it and saving it from the evils of excessive competition. During the depression of the 1870's, when he acquired most of his refineries, he considered Standard "an angel of mercy" rescuing improvident businessmen from ruin. Embittered oilmen went to the opposite extreme in appraising this type of salvation. Their thinking was epitomized by one small producer who, stubbing his toe on a rock, said vehemently, "Damn Standard Oil!"

Chicanery alone could never have made Standard the greatest company in the world. Rockefeller practically invented efficiency. With his vision of how to operate on a grand scale, he cut the pattern for twentieth-century business. As Rockefeller absorbed companies he also absorbed brain power, hiring his most brilliant competitors. Everyone who joined him prospered far beyond his individualistic dreams.

However, Rockefeller did not realize that the bigger a business becomes, the more carefully it must be conducted with regard to public opinion and the public interest. He was contemptuous of both the public and the small businessman, and for the same reason. Neither, he believed, knew what was good for the country or business.

After he formed the great Standard Oil Trust in 1882, he proudly declared, "The day of combination is here to stay. Individualism has gone, never to return." Rockefeller had no understanding of the kind of American individualism that prompted small oil producers along the Clarion River in Pennsylvania to state: "Resolved, that in consideration of the position taken by the 'large producers' in the matter of the combination, the 'small producers' of Clarion County do hereby resolve that we, the said 'small producers' don't care a damn for the combination or for the 'large producers' or

for anybody else and that's what's the matter with Hannah."
What Rockefeller could not understand was that the American temperament would never care a damn about combination, and that's what would always be the matter with Hannah.

Following Standard's lead, one industry after another combined to form monopolistic trusts—whisky distillers, tobacco manufacturers, sugar refiners. Power by combination, as the best means to reap big profits, was the creed of the times. A wrathful public, awakening to its exploitation, demanded protection. Standard's size made it the chief target. The Sherman Anti-Trust Act of 1890 resulted, but a club hurts only in proportion to the strength with which it is wielded. The government was more sympathetic to big business than it was to public opinion, and the Act bounced off the capitalists' back like a rubber stage prop. They promptly found a legal way to continue their trusts—devising holding companies under New Jersey's corporation laws. Standard again set the precedent. The end of the business depression in the mid-nineties and the diversion of the Spanish-American War muffled cries for dissolution. The antitrust club was to gather dust until Teddy Roosevelt's strong arm picked it up. The dust was so thick in 1900 that not a single suit was filed that year against 187 flourishing trusts. The industrial buccaneers who had raked the country's wealth into great glittering heaps were convinced that their control would never be seriously challenged. Now, with the notable exception of Rockefeller, they turned to the enjoyable task of flaunting their riches.

The way in which they did it awed even a member of the Russian court, fabled for its opulence. After watching this new American society acting the peacock, Grand Duke

Boris, brother-in-law of Czar Nicholas II, marveled, "I have never dreamed of such luxury . . . we have nothing to equal it in Russia."

The sumptuous parade fascinated rather than repelled less fortunate Americans. The nation's social conscience was as yet only muttering in its sleep. People were generally indifferent to the economic origins and implications of this startling pageant. The show was the thing.

The society pages of the New York papers were to grown-ups what Hans Christian Andersen was to their children. What writer of fairy tales could draw from imagination a more vicariously thrilling description of a princess than the society reporter of the *Times* could draw from his view of life at the opera?—"In Box Seven was Mrs. Astor who wore black velvet, the bodice smothered in white lace, elbow sleeves with flaring edges trimmed with lace and ropes of diamonds draped on the bodice and a band of black velvet studded with diamonds around her neck. About her waist was a loosely woven golden girdle studded with diamonds. She wore a diamond tiara and she carried an immense bouquet of white roses."

Only a few appreciated the caustic comment of the humorous magazine, *Life:* "And in her pockets? Diamonds? Very likely. And isn't it just lovely to feel that Mrs. Astor can do it! Think of saying to a servant, 'John, go down cellar and bring me a pint of diamonds from the second barrel on the left.' "

Gustavus Myers, a young political reformer and writer, canvassed New York's publishing houses to see if one would accept a book telling the truth about how the great American fortunes had been made; how the Astor fortune was founded on debauching Indians in the fur trade, shady real-estate

deals, and sharp practice in foreclosing mortgaged farms in
Manhattan; how Vanderbilt, Gould, Harriman, Fisk, Drew,
Rockefeller, Morgan, and others had amassed their wealth
by extortion, bribery, fraud, blackmail, manipulation of
stock, looting of companies, secret rebates, forcing competitors to ruin.

No publisher was willing to risk the powerful wrath of an
exposed plutocracy. If Myers would write a book in the
popular trend romanticizing and eulogizing their careers, em-
phasizing that their riches resulted from thrift and "sagacious
ability," it might prove a best seller. Myers, with the courage of a David, preferred to stick to his slingshot. It would
not be until 1909 that an equally courageous Chicago publisher would bring out Myers' classic *History of the Great
American Fortunes*. Meantime, unexpected by plutocracy,
public, Myers, or publishers, a series of basic discoveries and
inventions were about to mark the end of an era and the
opening of a spectacular new one in which Astors and Vanderbilts would be as effectively relegated to the status of
curiosities as their counterparts, the dinosaurs, were in geologic history. Only John D. Rockefeller would manage to
profit even more by the course of events.

On January 13, 1901, there was brief mention on the New
York *Times'* front page of the first of these discoveries, but as
it was tucked away at the bottom of the day's really important story, it is doubtful that even those with eyes to see
appreciated its significance. How could a few paragraphs
about an oil strike south of Beaumont, Texas, compare in
interest with column after column of glamorous reading
about the bridesmaids' dinner of the biggest society wedding
of the day? Even though the Texas dispatch called it "the

greatest oil strike in the history of that industry," surely this was not so interesting as the romance of one of the richest and most eligible bridegrooms in American social history?

Miss Elsie French, of New York and Newport, had captured Alfred Gwynne Vanderbilt, who had just inherited some $60 million, part of the original railroad fortune garnered by his great grandfather, Cornelius Vanderbilt. The delicious course of the attendant rites was charted daily on front pages in such detail as today is accorded meetings of heads of states. Sighing over the final ecstasies of the wedding as reported in the *Times* a few days later, most readers probably overlooked the paragraph on the same page advising that oil producers were rushing to sell their product to Standard. Prices had dropped 5 cents a barrel. "No bomb," continued the story, "has been thrown into the oil trade in late years that caused such wide consternation as the announcement of Col. J. M. Guffey's great strike in southeast Texas."

It was unlikely that John D. Rockefeller, reading his morning paper, shared in this "consternation" any more than he was interested in the French-Vanderbilt frivolities. He might have smiled thinly at the idea of any "bomb" being thrown into the trade he had so thoroughly organized and monopolized that after thirty-six years of shrewd, unremitting labor he was perhaps the richest man in the world. He could not have dreamed that the Texas strike was indeed a bomb—part of a chain reaction that would blow him into the billionaire class. Paradoxically, this elderly, unemotional man, who believed in secrecy and anonymity as though they were the eleventh and twelfth commandments, would presently find himself in the center of the stage of public opinion, there to be booed and hissed more bitterly than any American businessman, before or since, has ever been.

As for weddings, Rockefeller's New York town house was bustling with its own preparations. His daughter Alta was to be married to a Chicago attorney, E. Parmalee Prentice, four days after the French-Vanderbilt affair. New York society was no more aware of this than they were of how really rich the Rockefellers were. It is doubtful that even the Vanderbilts, whose seven palaces on Fifth Avenue had been built at a cost of over $12 million, realized that the family who lived around the corner in an ugly, narrow, four-story brownstone house, bought furnished for a modest $600,000, was already worth more than two generations of Vanderbilts had amassed.

The Rockefellers were just not society, and the wedding of one of the richest heiresses of the time rated only five paragraphs in the social column of the New York *Times* the day after the event. The trouble was, the Rockefellers did not observe the rule of the game. To be society, you not only had to be rich, you had to be constantly proving it. Unfortunately, Rockefeller's dollars did not leap lightly out of his pockets. His concept of spending was such that when he took his family to Europe, they checked every hotel bill in detail. At one meal, when a family discussion arose as to whether they had been charged correctly for two chickens, Rockefeller settled it by asking each diner to remember if he had eaten a chicken leg. Three had been eaten, so he was satisfied that he had not been overcharged.

Business, to Rockefeller, was the meaning of life, and his dedication to it resembled a great spiritual leader's dedication to his religion. He celebrated the anniversary of his first job with more ceremony than other people give to their birthdays, raising a flag in front of his home and inviting guests to help him honor the occasion.

To this man, the discovery of oil in Texas in 1901 was less important than the next move he would make in his chess game with J. Pierpont Morgan. Morgan, the great banker, had manipulated his pawns into position to form the biggest business merger the world had ever seen. He was planning the United States Steel Corporation, intending to control nearly all the iron-ore and steel manufacturing interests in the country. Only Rockefeller, who owned the world's richest iron mines, the Mesabi range, and dominated Great Lakes shipping, could checkmate Morgan's power game of a trust to control everything from raw material to finished product.

Morgan, Rockefeller, and the railroad combine already affected every aspect of American economic life to such an extent that there was a popular joke about it: "Who made the world, Charles?" "God made the world in 4004 B.C., but it was reorganized in 1901 by James J. Hill, J. Pierpont Morgan, and John D. Rockefeller."

The princely Morgan and the ascetic Rockefeller disliked each other intensely. However, Rockefeller, as the originator of industrial concentration, could approve and admire Morgan's plan. He sold his holdings to Morgan, pocketing another $50 million for what was actually a side interest.

When the formation of United States Steel, the world's first billion-dollar company, was announced, the dismayed public could scarcely laugh at the idea of Morgan and Rockefeller reorganizing the world. The trusts were moving toward complete control of the nation, and Rockefeller might be right. Individualism, if not yet gone never to return, seemed to be on its way out.

Fortunately, throughout America so many people were so busy being individualistic they had no idea they were supposed to be living on borrowed time.

In January, 1901, more than half the country's 76 million people were living on farms with no more freedom of movement than a horse could give them on roads still as primitive as those the founding fathers jolted over. Wagons bogged down in a winter's mire were not pulled out until spring. In a town of 5,000 people, all business would be virtually suspended after a heavy rain, as there were no pavements or sidewalks except in big cities. The nation's railroad system was practically complete, but the rigid network of rails froze movements of goods and people into an arbitrary pattern that did not permit great economic and social expansion.

As the new century began there were 4,000 horseless carriages, powered by steam, electricity, or gasoline engines, but they were still just playthings for the rich. A popular jingle summed it up:

> 'Twas said by a Whig
> That a man with a gig
> Enjoyed a clear claim to gentility,
> But a man who would now
> Win a parvenu's bow
> Must belong to the automobility.

As a spectator sport, horseless-carriage races were becoming increasingly popular, but even the government could not see this new machine as anything to be counted on for practical use. In 1900, the War Department had three automobiles specially equipped so that mules could be hitched to them if they didn't run. Only bicycles, manufactured at the rate of a million a year, offered freedom of movement independent of the horse.

For the majority of Americans of 1901 the distance between one another was almost as great as the vast gulf between rich and poor. Millions of immigrants were arriving from Austria, Italy, Russia, Germany, the British Isles, providing manpower and impetus for factories mushrooming around the big cities, for the coal mines and blast furnaces that were beating out a new, faster rhythm for the nation. These new citizens were trapped in industrialized centers, unaware of America's beauty, fruitfulness, and spaciousness. Its people leading separate lives everywhere, cut off physically from each other, though part of a common culture, a great, curiously poverty-stricken, heterogeneous nation was about to be quickly fused into an astonishingly close-knit, prosperous unity.

On farms and in small towns certain young men were having visions that would have seemed like hallucinations to the aging plutocrats in their New York palaces, who believed they had reorganized the world into its best and final shape. None of these earnest young men were aware of what the others were dreaming or that the fulfillment of each vision depended on that of the other. The simultaneous discoveries and inventions of these young men would do more for the complete change and advancement of American industry during the first fifteen years of the twentieth century than had been done during the whole century before. They would give Americans freedom of movement and create opportunities for wealth for the many instead of the few.

Oddly enough, some of these dreams and visions seemed about to fade away. In January, 1901, Detroit newspapers carried a dissolution notice of the Detroit Automobile Company with sale of all its assets. Henry Ford, Michigan farmboy turned mechanic, had dreamed of quantity production and failed. Nobody wanted his ridiculous-looking horseless delivery wagon. Broke and discouraged, Ford toyed with the idea of building a racer, since such a vehicle was apparently the only sort that could capture public interest.

In Dayton, Ohio, two youthful bicycle repairmen, Wilbur and Orville Wright, were equally discouraged. They had just returned to their bicycle business from a venture at Kitty Hawk, North Carolina. Their flight of 389 feet on that windy beach had broken the glider distance record, but they couldn't see a cent of profit in what they were doing or any notable contribution to human knowledge. They were debating whether it was worth it to go back and try again the next summer. "Man won't be flying for a thousand years," Wilbur concluded in disgust.

That victory for both Ford and the Wrights would come within two years seemed as unlikely as the fact that the basis for their success lay in Texas, where another young man— an Austrian immigrant—had already fulfilled his own vision.

BOOK TWO

Prodigal
Provision

*The misery of man appears like
childish petulance, when we
explore the steady and prodigal
provision that has been made for
his support and delight on this
green ball which floats him
through the heavens.*

RALPH WALDO EMERSON

1 : *That oil strike south of Beaumont*

Big hill, as the natives called it, was a low circular mound, about half a mile in diameter, rising no more than twelve feet above the surrounding swampy plain. The Neches River curled lazily nearby, emptying its waters and barges of rice and lumber into the Gulf of Mexico a few miles further south. There was not much to distinguish Big Hill from any other low mound anywhere except that near its crest there was the rotten odor of sulphurous gas seeping from the soil. Venturesome schoolboys from Beaumont found rich satisfaction in lighting the gas and watching the earth burn.

Every curious thing in nature attracts to it a curious mind. Big Hill attracted Patillo Higgins, the one-armed, lighthearted, self-educated, jack-of-all-trades son of the local gunsmith. In the late 1880's, hoping to promote a brickyard in Beaumont, Higgins had traveled to Pennsylvania, Indiana, Ohio, and New York to find out how brickyards operated. Learning they used oil and gas as the most efficient fuel, he also visited several oil fields. There he found the drillers following the only clues nature gave them, oil and gas seeping up in springs. Higgins recalled Big Hill's gas. If in Pennsylvania, why not in Texas, too?

Higgins dreamed big. He visualized thousands of barrels of oil coming from the mound's center, the rest covered by a new industrial city. He saw Beaumont as a great inland port. So convincing was his picture that he persuaded the town's leading citizens to finance the Gladys City Oil, Gas and Manufacturing Company, named after a little girl in the Sunday-school class he taught. The company bought and leased most of the land on and around the mound, and Hig-

23

gins prepared an elaborate plat for the new city, including locations of hotels, churches, schools, parks, fire stations, a hospital, and even a college. All that remained was to drill a well down 1,000 feet to where Higgins said the greatest oil pool in the world was.

It wasn't quite so simple. The first attempt went only a little over 400 feet. The earth beneath the mound was unlike the Pennsylvania rocks, where a hole could be hammered down by the crash, lift, crash, lift of heavy iron tools. The mound was made of heaving quicksands through which there seemed no way to make a proper hole. Two more attempts could not get even as deep as the first.

A year after Higgins' attempts, in 1894, the city of Corsicana, 200 miles northwest, needed water. It drilled 1,000 feet and found oil. The ensuing flurry of drilling didn't find big quantities of oil in any one well, but produced a respectable total of over 2,000 barrels a day. Texas had become an oil state without Patillo Higgins' help.

After seven frustrating years Higgins had exhausted his own financial resources and those of his backers. His insistence that there was oil in Big Hill was now a local joke. Beaumonters put their faith and money into such ventures as furniture factories and rice mills. You could see the raw material for these on top of the ground. The quiet little town was Southern in spirit and its citizens were content with the slow, orderly rhythm of the established businesses along the magnolia-shaded streets.

Hoping for scientific support of his idea in order to raise more money, Higgins arranged for a state geologist to visit the mound. The petroleum geology of the 1890's was no more scientific than fortune telling; Higgins would have done

as well to get a gypsy. The geologist could find no reason
why oil should exist in the coastal plains, and to Higgins'
horror he published an article in the town's only newspaper
warning everyone "not to fritter away their dollars in the
vain outlook for oil in the Beaumont area."

Higgins remembered feelingly the outburst of Governor
O. M. Roberts fourteen years earlier. The governor had appointed a committee to investigate the possible resources of
oil, minerals, and artesian water in Texas. After a six-month
survey the committee reported that there were no prospects
or indications of any of these resources in the entire state.
"Gentlemen, as governor of the state of Texas I accept your
report and discharge the committee," Roberts said indignantly. "But I want to say to you that every word in your
report is a lie. I repeat it, a lie! Do you think God Almighty
would create a great country like this and not provide a way
to take care of it? In the very nature of things your report
is a lie!"

The find at Corsicana had vindicated Governor Roberts.
Now it was up to Higgins to prove his faith. In desperation
he advertised in an Eastern trade journal. There was only
one answer, but that was enough. It was from Captain
Anthony F. Lucas.

To Big Hill, waiting like a princess for her deliverer, the
Captain was Prince Charming. He had the imagination and
vision of a dreamer, the mind and education of a scientist,
the daring of a gambler, and, to Patillo Higgins' relief, money
to invest. Only the tritest of historical romancers would have
invented such a handsome, gifted adventurer to play the
hero's role. Of Slavic origin, the Captain had graduated as
an engineer from the Polytechnic Institute at Gratz, Austria,

and been commissioned in the Austrian navy. After visiting an uncle who had emigrated to Michigan, he became an American citizen, changing his name from Luchich to Lucas.

His career seemed a miscellany of unrelated experiences, but each one had sharpened a particular talent or added new vision, all of which was essential for him as he faced the challenge of his great opportunity. In a Michigan sawmill he developed his ability as a mechanical engineer, redesigning a gangsaw. He caught the exploration fever as a gold prospector in Colorado. He married the right girl, a Georgia doctor's daughter, who was to give him what he needed most at a critical moment—courage to go ahead. The next step was an odd one—a job as engineer in an island salt mine in a Louisiana bayou.

The mine was "a weird and beautiful cavern. Arch after arch stretches away, and looking down the dark and gloomy avenue one is amazed by the inexhaustible deposit, and when it is artificially lighted up millions of crystals flash and sparkle with wondrous splendor." So wrote Joseph Jefferson, who had immortalized the role of Rip Van Winkle on the American stage. Living in retirement on a magnolia-covered island near the mine, Jefferson had found salt when he drilled for water. The arrival of the mining engineer gave him an idea. Perhaps he, too, possessed an "inexhaustible deposit" that would reflect a certain wondrous splendor in his bank account.

Captain Lucas drilled for Jefferson finding pure rock salt at 350 feet. The well was still in salt at 2,100 feet when they stopped. Unfortunately, there was also water, ruining the chances for a mine. Lucas and Jefferson would never know how close they had come to finding a great oil pool on the island. Another well would discover it forty-two years later.

The Captain now had experience in drilling, and on his own account began exploring other great salt masses in the bayous. On two he found some oil and sulphur associated with the salt. He began to reflect. Neither he nor anyone else knew that these masses were great plugs of salt mushrooming up from a mother salt bed formed in dry, hot times over 200 million years before, in some places lying as deep as 45,000 feet. The salt, lighter than its burden of sediments, had slowly pushed its way up in hundreds of spots along the Gulf Coast, doming the surface of the earth over the plugs. As the plugs thrust their way upward, they tilted the layers of sands, shales, and limestones so that traps formed into which vast quantities of oil migrated.

Captain Lucas perceived that oil, sulphur, salt, and mounds were somehow related. When he read Patillo Higgins' advertisement telling of a mound with gas and sulphur that Higgins believed would produce oil, he immediately suspected the presence of a salt dome, although there was no mention of salt. Like Higgins, but for different reasons, Lucas began to believe in Big Hill. He took a lease from the old Gladys City company on about six hundred acres on the hill, paying the company $11,050 and signing notes to pay $20,100 more. To Patillo Higgins went a 10 per cent interest in the Lucas lease.

On a hot July day in 1899, the Captain confidently started drilling with equipment he had brought from Louisiana. He expected to finish in a few weeks, but six months later he had managed to fight his way down through only 575 feet of the quicksands. However, the hill seemed to be trying to encourage him, for a little heavy green oil came out of the hole. Then gas pressure collapsed the pipe in the hole and his inadequate equipment could go no further. He couldn't even start a new well, for like many an oilman before and after him, all his capital had been drained into an abandoned hole in the ground.

Money, after all, is only evidence of the resources of the spirit, and in these both the Captain and his wife were rich. Her faith matched his, and she encouraged him to seek financing. The Captain approached Standard. Standard sent its production expert, Calvin Payne, to examine the mound. Payne had been in the oil fields of Russia, Rumania, Borneo, and Sumatra as well as all the fields in the United States. He reported to his company that there was "no indication whatever to warrant the expectation of an oil field on the prairies of southeastern Texas." The mound, he said, had "no analogy to any oil field known, as in fact there was not the slightest trace of even an oil escape." He also gave the Captain some free advice: Go back to mining and stop wasting time looking for oil.

Dr. C. Willard Hays, soon to become chief of the United States Geological Survey, happened to visit Beaumont. The Captain eagerly called on him, sure that the famous scientist would understand and corroborate his deduction that oil accumulated around salt masses. But Dr. Hays was no more perceptive than Standard's Calvin Payne or the state geologist who earlier had dealt Patillo Higgins such a crippling blow. Hays could see "no precedent for expecting to find oil in the great unconsolidated sands and clays of the Coastal Plain," which simply meant that, like Payne, if he couldn't see any seeps there wasn't any oil.

"The plain fact of the matter is," Lucas sadly concluded, "that I am not a trained geologist hence do not see my way to give the proper or necessary interpretation to my—well—visions."

There was one geologist who believed in the Captain. As in Higgins' case, one believer was enough. Dr. William Battle Phillips, a University of Texas field geology professor, arrived in Beaumont, like a messenger from Divine Providence, to tell Lucas that although he could not afford to take issue with such eminent federal and state authorities, he thought Lucas' theory was right and would be glad to give him a letter of introduction to a Pittsburgh oilman, John H. Galey, who might be interested in drilling the mound.

Galey and his partner, Col. James M. Guffey, were nationally known oil prospectors. Guffey and Galey—they sounded like a vaudeville team, and Guffey looked and acted the part. A politician and money raiser, he affected wide-collared, pleated shirts, a Windsor tie, dazzling waistcoats, and a Prince Albert coat. His mane of curly snow-white hair was always topped with a wide-brimmed black felt hat. A flowing white mustache emphasized his delicate features and complexion. He had been chairman of the Democratic National Committee during President Garfield's time and had the appropriate personality and ambitions.

Galey was the oil finder of the team. Slight, modest, inconspicuous, he loved prospecting for the game itself. He and Guffey had been highly successful in Pennsylvania and Kansas, developing properties and selling them to Standard. When oil was found in Corsicana in 1894, Guffey and Galey were the first Eastern oilmen on the scene.

Galey caught Lucas' vision. This was enough for Guffey; he promptly started trading. He dwelt on how Standard had turned down the deal, how the prospect had been condemned by United States Geological Survey and Texas geologists, how four dry holes had already been drilled. The 600 acres under lease weren't enough to make an exploration program worth while, Guffey said. If Lucas would lease all the acreage on the mound and around it, then Guffey and Galey would consider investing $300,000, giving the Captain an eighth interest in the prospect. If the first well did not find oil or favorable indications, then they would not be obligated to drill more. The Captain was no match for the Colonel by training or nature. He accepted the partners' offer, proudly insisting he be paid no salary nor be reimbursed for his own investment. This gesture would soon mean that in order to eat while the well was drilling, Mrs. Lucas would cheerfully sell their furniture and use egg crates and apple boxes for chairs and tables.

The Captain managed to lease about fifteen thousand acres on and around the hill. Colonel Guffey immediately borrowed $300,000 from Andrew W. Mellon and Big Hill was in business again. It could only have been the attraction of opposites that enabled Guffey to cast an almost magical spell over Mellon, the shy, quiet little Pittsburgh banker who was already on his way to becoming one of America's four billionaires. Mellon's advisors and partners disliked and distrusted the Colonel, but the Colonel could always borrow money from Andrew himself. Andrew genuinely liked him. Besides, all the Mellons had complete confidence in what W. L. Mellon, Andrew's nephew and the oilman of the family, called Galey's "amazing power to scent a hidden pool of oil underground. Any time John Galey was convinced that there was oil some place it was worth a chance."

Galey picked the spot for the fifth try on Big Hill—a wallow where farmers brought their hogs to let the sulphur water kill their fleas and cure their mange. A nose less sensitive to the smell of oil and more so to that of hogs might have staked the location 50 feet south and gotten a dry hole, as someone did later.

Galey called the Hamill brothers, Al and Curt, from Corsicana to drill the well under Captain Lucas' supervision. The Hamills used a rotary rig much like the one Lucas had used in Louisiana. By the time the three of them had finished the fifth well at Big Hill they had discovered and perfected the most important techniques of modern drilling for oil. They had to.

In the Eastern oil regions rotary drilling was almost unknown. There they used cable tools. The difference between the two methods was somewhat like the difference between driving a nail or putting in a screw. Cable tools pounded a hole. Rotary drilling had been devised as a faster means of going through soft formations for salt and water. A "fishtail" bit was attached to pipe and the pipe was rotated mechanically from the surface, the bit gouging its way down. Water pumped down the hollow drill pipe kept the bit cool and forced the cuttings up the outside of the pipe. The method was ideal for drilling in the soft formations of Texas and was in general use in the Corsicana field.

The Hamills, having to master Big Hill's baffling quicksands, made an important discovery. If, instead of water, they pumped mud down their drill pipe, it plastered the sides of the hole, making it firm. They drove a herd of cattle through their water pit to make thicker mud. From this simple beginning a multimillion-dollar industry grew to supply specially mixed drilling muds to cope with similar problems in other types of formations.

At 160 feet they drilled into a layer of coarse sand and gravel that even mud couldn't seal off. Al Hamill was fond of saying later that it simply took a strong back to drill this well. The brothers now performed an extraordinary physical feat. They could think of no way to take the hole down except to drive the 8-inch pipe by hand and wash up the debris inside it, so they rigged up a heavy block that, when lifted by a line and quickly let go, would strike the pipe, driving its beveled cutting edge a little further into the rocks below. "We took turns on this man-killing job," Al Hamill said simply. It took two weeks to go through this formation, which was 285 feet thick. The turns were longer and harder after one of the Hamills' two helpers quit in disgust.

Big Hill wasn't through testing them. After the relief of a few hundred feet of normal drilling they hit a gas pocket that forced sand and mud up the pipe. They couldn't screw on additional lengths of pipe, for when the counter pressure of the mud pump was stopped the pipe would stick. It was impossible to drill further.

Captain Lucas' mechanical-engineering mind couldn't sleep that night. Inspiration finally came, as it does so often, out of the dark silence. A back-pressure valve could be put in the pipe on the same principle by which water is pumped into a steam boiler without allowing the steam to escape. Elatedly, Lucas fashioned such a valve the next day. Drilling resumed, and his invention has since become an indispensable part of all rotary rigs.

In six weeks they had driven the hole down 300 feet beyond where the Captain had been forced to abandon his first well after six months' labor. At three o'clock on a cold mid-December morning, Al Hamill was working his lonely 18-hour shift. The drilling became easier. The bit went faster and faster. As the sky lightened, Al's heart caught the drilling rhythm for he could detect a little oil in the slush pit where mud from the bottom of the hole poured out. By the time his brother and their helper arrived with his breakfast there was a big showing of oil.

When Captain Lucas arrived, he excitedly asked Al how big a well he thought it would make. Al's only experience had been with Corsicana's small 10- and 25-barrel wells, so he named a daring figure. "I think it will easily make fifty barrels a day."

The Captain wired Galey. The wily old prospector was jubilant, but instead of completing the well, he decided they should go on down another 300 feet as planned. They could always come back to the showing of oil and produce it. But Christmas was at hand and the drillers certainly deserved a holiday. Big Hill slumbered on.

On New Year's Day, 1901, the Hamills were back on the job. In a week they drilled another 140 feet. At 1,020 feet they hit a crevice and the pipe turned helplessly round and round, making no progress. Al wired Corsicana for a new fishtail bit. He met the freight train early in the morning, January 10, and drove his buggy back to the well. With the new bit on, they had lowered about 700 feet of the drill pipe down the hole when suddenly the well began to spout mud. It drenched Curt, perched high on the derrick, and he scrambled blindly down. Then they watched in terror as four tons of heavy pipe shot up out of the hole, over the top of the derrick, the joints breaking as it went skyward. Lengths of steel pipe crashed dangerously around them, and then all was quiet again. The brothers and their helper ventured back on the rig floor and looked with disgust at the mess of mud and ruined equipment. Resignedly, they started shoveling mud. Then, with no warning, there was a deafening roar and the well erupted like a volcano: first mud, then gas, oil, and rocks shooting hundreds of feet in the· air.

It took the dazed men a few minutes to realize what was happening, for no man in America had seen such a sight before. Curt remembered the fire in the boiler. "We've got to get that fire out!" he yelled. With oil showering them they threw buckets of water on the fire box until the last spark was doused.

The helper went to find Captain Lucas. When Lucas came up over the hill his horse was at full run, but the Captain needed wings. He jumped from the buggy, picked himself up and ran on, not daring to believe his eyes, shouting, "Al! Al! What is it?" "Oil, Captain—it's oil!" "Thank God!" cried Lucas, grabbing the driller and hugging him again and again in his excitement.

The roar had shaken the countryside and the black geyser could be seen for miles. Crowds of ranchers and their wives began arriving on horseback, in buggies, on foot. A rancher rode into Beaumont, shouting "They've got a wild oil well out at the hill and the damn thing's ruinin' my land!" Within a few minutes the town was empty, everyone heading for the hill. Hypnotized, the crowd watched the well gush. A dangerous lake of oil started forming. The Hamills recruited watchmen from the crowd to prevent people from smoking.

Patillo Higgins was among the last to know. He had left town early that morning to see about a lumber deal that might get him out of debt. No one mentioned his name in all the excitement. Captain Lucas was the deliverer of Big Hill and the hero of the day. Higgins' name would even be absent from the towering granite shaft erected forty years later, commemorating the site of "the first great oil well in

the world—the Anthony F. Lucas gusher." But Higgins unwittingly gave the whole oil field its popular name. While
the well was drilling, he bought 33 acres on the hill from a
real-estate company that called it Spindletop Heights. When
he told newspaper reporters he was going to drill a well
there, the name caught their fancy. There were other mounds
called Big Hill. The oil strike south of Beaumont needed a
truly different name to distinguish it. Spindletop was just the
thing.

2 : *"The cow was milked too hard"*

THE GREAT WELL was gushing 100,000 barrels a day in a
solid black stream 175 feet into the air. Texas railroads advertised "In Beaumont, You'll See a Gusher Gushing," and
within two days brought 10,000 sightseers to marvel at the
phenomenon. San Franciscans, reading in their *Chronicle*
that Beaumont was "oil wild and land crazy," might have
thought they knew from experience what was going to
happen now, but the California gold rushes were like summer thunder showers compared to the tornado boom that
would soon devastate Beaumont.

At the height of the Gold Rushes anyone could pan ten to
fifteen dollars' worth of gold dust by working from dawn to
dusk. Exceptionally good digging might sweat you out $500
in a day. In the oil-boom lunacy no real work was needed to
make incredible profits. A St. Louis *Post-Dispatch* reporter,
stepping off the train into the Beaumont madhouse, was
offered a lease for $1,000. He laughingly refused, but a St.
Louis man in back of him bought it. Two hours later he sold

it to someone else for $5,000, only to see it sold almost immediately for $20,000. No one knows how much money changed hands in the frenzied trading that lasted for almost a year in this oil-struck town. The combined gold rushes of 1848 and 1849 produced only $18 million worth of gold. When the Beaumont boom was at its peak an acre of land was sold for $1 million and none for less than $200,000. Within a year the land at Spindletop was assessed at $100 million and the capitalization of the oil companies in the field had reached $200 million.

The boom did not really get under way until April, 1901. Lucas' well was so fantastic it was generally considered a freak of nature that could not possibly happen again. After nine days the Hamill brothers were able to cap the gusher's flow and then there was nothing to attract sightseers except the vast shallow lake of almost a million barrels of oil around the sand mound covering the spot where the well had been drilled.

The Mellons, too, were reserving judgment on the significance of the gusher. Colonel Guffey was urging them to put up the millions necessary to finance a big oil company. There were no refineries or pipelines in Texas, nor as yet any market for the oil. The bankers were not willing to commit great outlays of capital until it could be proved that the discovery wasn't a freak. So Galey and Guffey began drilling two more wells. D. R. Beatty, an imaginative Galveston real-estate promoter, arriving the day after the Lucas well blew in, had leased 10 acres in Spindletop Heights near Patillo Higgins' land. Now he was also drilling a well. Scott Heywood, an ex-vaudeville performer and Klondike gold-rusher who had dabbled in California oil, grabbed the first

train from San Francisco when he read the news of the
"spouter." He leased 15 acres from Higgins and he, too, had a
well going down.

On March 3, sparks from a passing train set fire to Lucas'
lake of oil. The blaze was a "spectacle of unparalleled
grandeur" to Lucas even in the midst of frantic attempts to
put it out. "When we found there was no possible hope of
saving the oil we started a counter-fire about a mile below
the oil lake," he wrote. "When the two conflagrations met
there was a heavy explosion which threw the blazing oil high
into the air while the earth trembled as if shaken by an
earthquake. We were glad to know that our great well was
perfectly safe having been covered with sand in view of this
very contingency."

Scarcely had the fire died down when D. R. Beatty's well
roared in, spouting almost as high as the Lucas well. In rapid
succession the Guffey and Galey, the Higgins, and Heywood
wells came gushing in. A startled world received the news
that six wells in Texas could produce as much in one day
as all the oil wells in the whole world.

Here was prodigal provision indeed. What's more, it was a
free-for-all. Captain Lucas was responsible for the ensuing
boom in more ways than one. When Guffey told him to get
more acreage, he felt justifiably proud of leasing 15,000 acres.
He concentrated on large landowners, since it took quantity
to satisfy Guffey. There were countless small owners, each
owning a few acres, whose land he did not bother to lease.
For speculators, this was Paradise, Valhalla, Nirvana, and
El Dorado rolled into one. Guffey, surprisingly enough, enlarged the happy hunting grounds by selling 15 acres on the
hill for $180,000 to a syndicate headed by the state's ex-

governor, Jim Hogg. The syndicate sold 2½ acres for $200,000 and then merrily chopped the rest into profitable pieces as small as ⅟₃₂ acre.

Forty thousand strangers descended on Beaumont's 10,000 residents. It was no longer a town. It was a seething, frantic concentration of fortune seekers, swindlers, gamblers, prostitutes, sellers and buyers of anything and everything. Saloons never closed. Men slept in barber chairs, on pool tables, in store windows. Three hundred cots in the town auditorium were rented in 6-hour shifts. The night train from Houston, 80 miles away, became a Beaumont hotel, with men buying 30-day tickets and catching an afternoon train from Beaumont to Houston in order to sleep on the night train back to Beaumont. There was little reality in values in this derangement. Oil was so plentiful it sold for 3 cents a barrel, drinking water so scarce it sold for 5 cents a cup. When doctors announced it was safer to drink whisky than Beaumont water, the local WCTU chapter indignantly dispensed free boiled water.

Men hawked oil leases on the streets like circus candyvendors. They made desks out of packing cases and opened offices on the sidewalks. The center of the fevered trading was the Crosby House lobby. Men stood on the hotel's chairs day and night auctioning off leases. An acre of land bought at $8 was sold and resold so many times in the crowd that the last buyer paid $35,000 for it. Pieces of land 25 by 34 feet sold for $6,000. Land 150 miles away from Beaumont sold for $1,000 an acre. The traders carried money around in suitcases, paying for $100,000 purchases in cash. There was such a need for cash that a principal shipment into Beaumont was freight cars loaded with silver dollars. Barbershops and restaurants routinely made change for $1,000 bills.

On the hill, wells were drilled so close to one another that a man could walk from one rig floor to another without ever touching the ground. Every well was a gusher. One was drilled on the spot where Captain Lucas had had to abandon his first well at 575 feet. The confident Hamills guaranteed gushers. If they drilled a well that didn't gush, no payment was required. The number of barrels a well produced became meaningless. The way to be sure a well was a gusher was to know the size of the pipe the oil flowed through. "R. A. Josey has sold to H. L. Fagin a six inch gusher in Block 38 of the Hogg Swayne Syndicate tract," ran a typical announcement in the Beaumont *Journal.*

By the year's end there were 440 gushers on the hill and the oil was pouring out as though through a sieve. Nobody knew that the field was being badly damaged so that millions of barrels would be irrecoverable. The gas pressure needed to bring the oil to the surface was being released all at once. When Captain Lucas visited the field three years later, he was saddened to see the devastation. Of the 1,000 wells that had been drilled, only 100 were producing so much as 10,000 barrels a day. "The cow was milked too hard," Lucas said, "and moreover she was not milked intelligently."

There was no sense to anything in the Beaumont boom. Nature was on the rampage and the boomers followed suit. Even the local garbage woman became a national celebrity. Nicknamed "Mrs. Slop," she leased her little pig farm on the hill for $35,000 and kept right on collecting garbage. A railway company employee bought a lease and sold it for $200,000 profit. The railroad naively issued an order forbidding employees to engage in such activities.

Sensational fires swept the hill, but derricks were rebuilt and more wells drilled. When heavy rains turned Beaumont

and the hill into a sea of liquid mud, the boomers used row-
boats.

Spindletop was a nationally advertised curiosity. Excur-
sion trains brought 15,000 people a day to see its wonders. If
a gusher wasn't being drilled in, nobody had to go home disappointed, for stock promoters would open up wells and
let them gush. Who could resist buying stock with such exciting proof before his eyes? The hill became known as
"Swindletop," and some men made a handsome living by
signing certificates for the hundreds of promotional companies that were picking the pockets of the whole nation.
The One Penny Oil Company, capitalized at $1 million, sold
stock for a penny a share. Another million-dollar company
was organized on a nonexistent lease. Stockholders in companies which had a lease were no better off. Not a barrel of
oil was being sold, since there was no market for it. One man
stated he "went busted drilling 10,000 barrel wells."

The companies had a curious assortment of owners. The
editor of the *Texas Baptist Standard* and the publisher of the
Texas Christian Advocate each had companies. The Drum-
mers Oil Company was formed with prominent traveling
salesmen as officials. The Young Ladies Oil Company, which
owned a half-interest in a "six inch gusher," had as directors
young ladies from Joplin, Fort Smith, and New York. The
citizens of Uvalde, Texas, organized a company to drill a
location selected by a boy with "X-ray eyes," who, blind-
folded, supposedly could see underground streams at night.

Although Spindletop would prove to be the first giant oil
field (the term for a field containing more than 100 million
barrels), there still wasn't enough oil in the mound to make
everybody rich who thought he was going to be. More
money was lost at Spindletop than was ever made. The

swindling was as inevitable a part of the boom as the brothels and the saloons and, like them, it was incidental to the really important events on the hill.

Colonel Guffey's third gusher convinced the Mellons. They and their friends financed the Guffey Petroleum Company, paying $1,500,000 for a half interest. The new company was to develop the Spindletop properties and build a pipeline and a refinery. Guffey shrewdly bought out his old partner and Captain Lucas. Although the bankers thought Galey would receive $750,000, the Colonel, knowing his partner's indifference to money, gave him half that much plus some miscellaneous stocks. Captain Lucas fared better. He received $400,000 cash, a thousand shares of stock in the new company, and a contract to explore for more salt domes. He left Beaumont to its boomers and soon retired modestly in Washington, D.C.

Patillo Higgins had also played out his role as oil finder. He profited even less than Lucas from Spindletop and lost his money in drilling other mounds that held no oil. When he died in the 1950's, however, he had no regrets. After all, there was only one Spindletop.

Colonel Guffey proved an indifferent businessman. After a year, when Spindletop's wells stopped gushing, the Mellons bought him out. "They throwed me out," the Colonel complained. In truth, the Mellons were trying to salvage their investment in a grandiosely conceived but inexpertly managed company.

The Mellons had ventured successfully in oil before. In the nineties Andrew's nephew, W. L., had developed production in the Eastern oil regions, selling holdings to Standard. Figuring it would take at least $15 million to tidy up Colonel Guffey's Spindletop mess and put it on a profit-

making basis, Andrew and W. L. tried to induce Standard to
take over. Rockefeller's right and left arms, John D. Archbold
and H. H. Rogers, were amused at the Mellons' predicament.
"After the way Mr. Rockefeller has been treated by the State
of Texas, he'll never put another dime down there," Rogers
said. Texas was so set against monopoly that it had outlawed Standard before a barrel of oil was discovered in the
state. Archbold and Rogers implied that the Mellons would
have Standard's blessing if they revised their plans and
stayed in the oil business. Since they could provide no competition, that is, they were to be tolerated.

Standard's attitude might not have been so benign could
it have known that the Mellons, under W. L.'s astute guid-
ance, would soon build one of the world's biggest oil companies, Gulf Oil Corporation. In a few years this child of
the Lucas gusher would be a major competitor of Standard.
Not even the Mellons could guess that Gulf would be their
own greatest money-maker, greater than their huge businesses of banking, aluminum, and carborundum and coke
processing.

However, Spindletop was lavish with its gifts. Standard
would benefit indirectly as much as all the others. Among the
thousands who crowded into the Beaumont carnival were
many serious, forward-looking young men. At Spindletop
they found inspiration for exciting, satisfying, and rewarding
careers. Two of them, W. S. Farish and Lee Blaffer, formed
the Humble Oil Company, which was to become a great independent oil-finding company. Eventually, when Standard
bought an interest in it, Humble would change Standard's
policy and history.

Spindletop gave birth to other big companies. The Texas
Company was organized by a group who bought producing

properties on the hill. The Sun Oil Company, a small independent refining company that had managed to compete with Standard in Philadelphia, saw its opportunity, bought properties, and began its rise.

The Beaumont boom was short-lived. Within six years, as at Babylon, there were "goblins howling in her palaces and jackals in her pleasant mansions." Only a few wells wearily pumped up the last trickles of oil from the top of the mound. However, within that short time the nation had become rich in a way the stock promoters would not have thought of promising. The flood of 50 million barrels of oil had swept away old industrial methods. Railroads, ships, and factories, using this cheaper, more efficient fuel, were multiplying as never before.

At the celebration of Spindletop's fiftieth anniversary, Dr. E. DeGolyer, one of the great oil finders, paid the hill its truest tribute. "I once traveled over the countryside of Cape Breton for some weeks," he said. "It was a hard land. 'What does this country produce?' I asked. The reply was from a dour Scot. He looked me squarely in the eye and replied 'Men.' And so it was with Spindletop. It was a producer of men."

Men from Spindletop forged a dynamic new art of oil prospecting. Inspired oil explorers were ready to go anywhere and everywhere now that Spindletop had shown there was an abundance of oil in the earth. Spindletop and its men proved there was something more colossal in America than its geography. Individualism gone never to return? It was just beginning!

CALIFORNIA HAD BEEN producing small amounts of oil since 1864, but years passed before the explorers knew they had found fields even bigger than Spindletop. There had been no gushers to indicate that within two years of Spindletop's discovery California would become the nation's biggest oil-producing state. However, like Spindletop, California's fields had produced men.

The first search in California was stimulated by Benjamin Silliman, the Yale University chemistry and geology professor whose report on Pennsylvania's rock oil had inspired the New Haven investors to drill the Drake discovery well. In 1864, Thomas A. Scott, Lincoln's Assistant Secretary of War and vice-president of the Pennsylvania Railroad, sent Silliman on a private mission to investigate stories about oil seeps in Southern California. Colonel Scott was interested in oil: he and Andrew Carnegie had taken a flier in the Pennsylvania boom, making a profit of $1 million by leasing the right farm.

"California will be found to have more oil in its soil than all the whales in the Pacific Ocean," Professor Silliman reported enthusiastically. "The oil is struggling to the surface at every available point and is running down the rivers for miles and miles."

With a railroader's dispatch, Colonel Scott had the first drilling equipment to reach the Pacific Coast on the way there before the last die-hard Confederate army surrendered in Louisiana. After drilling six wells and finding in only one of them a little heavy oil from which kerosene could not be extracted, Scott gave up in disgust, as did others who had excitedly followed his lead with no better results. California's first boom lasted a little over a year, and the explorers had spent $1 million to produce $10,000 worth of oil.

The subject was forgotten for the next decade as thousands of immigrants poured into the state from the embittered, ruined South. There were immigrants from the North, too, including veterans of the original Oil Region. These, try as they would to settle down as merchants, could not long resist the lure of the seeps and the promised rivers of oil.

In 1876, when an ex-Pennsylvanian finally drilled a well that flowed 150 barrels a day, the state's prospects were once again taken seriously. A number of small companies merged to form the million-dollar Pacific Coast Oil Company, backed by San Francisco capitalists. An official of this ambitious new company, visiting Pennsylvania, spoke glowingly to a boyhood friend, Lyman Stewart, who had become an oil producer. Why, California was already producing 40,000 barrels of oil a year and selling its refined products as far afield as Mexico and Hawaii. And this was only a beginning. If Stewart would come west, Pacific Coast Oil would lease him all the promising land he wanted.

Lyman Stewart was one of those strenuous souls who are beyond cheap success. When Edwin Drake made his discovery in 1859, the nineteen-year-old Stewart, a tanner's son in a nearby village, caught the fever. He invested and lost the $125 he had saved to become a missionary. With a fervor kin to that of saving souls he saved more money, only to lose it in a second venture. After three years as a Union Army volunteer, he was at it again, this time making a fortune of $300,000 in trading leases and buying small interests in wells.

Lady Luck seems to take revenge on all great prospectors who are unfaithful to her. Almost without exception, whenever they attempt to take their winnings from the gaming table to invest in something "secure," they lose with astounding rapidity. A mowing-machine factory took all Stewart's winnings. So it was back to work again to save more money for a new oil venture.

The rewards were not so great this time, for there was already an overproduction of oil in Pennsylvania and Standard Oil had monopolized its outlet. Stewart listened eagerly to his friend's dazzling stories of California. His partner, W. L. Hardison, was not so enthusiastic. They agreed to sell their properties and divide their profit. Stewart started across the continent with a stake of $70,000.

Seeing's believing, and in California the cautious Stewart became as confirmed a believer as Professor Silliman. The rivers of oil would really flow, he thought, if the wells were only drilled a little deeper. He wired his former partner with such conviction that Hardison soon arrived with two drilling rigs and experienced crews to operate them.

Though seeing may be believing, the forgotten part of the proverb is "but feeling's the truth." The truth, as so often happens, was elusive. After drilling six dry holes in a row the partners had used all their capital and were $183,000 in debt. In desperation, they persuaded Pacific Coast Oil to let them drill a proven lease. Finding some oil, they sold the property to pay their debts. Their luck had changed, but not enough. They were caught in a vicious circle. They would find a little oil—enough to borrow more money to drill more dry holes to put them deeper into debt. Borrowing still more money, they would find a little more oil, borrow still more money, drill still more dry holes. It was an agonizing way to try to make a fortune, but one all too familiar to thousands of oil explorers then and later.

Finding California's oil was so tricky that at one point the partners decided to try upside-down wells. A likely mountain having defied efforts to perch a drilling rig on it, they tunneled upwards from its base. A few barrels of oil trickled down the holes into tanks at their entrances; this slow bleeding has continued almost seventy years.

Although Stewart and Hardison were usually on the verge of financial ruin, they gradually expanded their activities to include refining, pipelines, and marketing, at length forming the Union Oil of California. The business hardly seemed worth the incessant hardships and drudgeries until one day in 1892, in a canyon in Ventura County, northwest of Los Angeles, one of Stewart's exploration wells, like a fountain, flowed 1,500 barrels of oil a day over the derrick, down the canyon into the Santa Clara River, and out into the ocean.

Lyman Stewart had fulfilled Professor Silliman's visions. Man, as well as Nature, could coat the rivers with oil, and the idea that there was more oil in California's soil than in all the Pacific's whales would never be doubted again.

Stewart could imagine no greater compensation for his hard, lean years, but he was to have one. In 1910, a group of Californians came to him for help. They had formed the Lake View Oil Company to drill in the San Joaquin Valley, and for a year had struggled to put down a hole with inadequate equipment and insufficient money. At 1,800 feet they had only hope left. It was not a particularly promising area, but Stewart, remembering his own painful struggles and unable to resist the lure of an exploratory well, agreed to complete the test, using Union drilling crews whenever they had spare time. In return, Union would receive 51 per cent of Lake View Oil's stock.

The odd-job well was assigned to Charles Lewis Woods, an

experienced driller who had earned the unlucky nickname of Dry Hole Charlie by never bringing in a producer. Drilling was continued on and off down to 2,200 feet, at which point work was stopped, awaiting the services of the next spare crew. On March 14, 1910, a nearby well in which Union had keener interest shut down temporarily, and its midnight crew was sent over to drill ahead on the Lake View well. When they pulled the bailer out of the hole they were amazed to find it dripping with oil. Oil started filling the hole, but no one was prepared for what was about to happen. The well seemed to be quietly waiting for Dry Hole Charlie to come to work at eight o'clock. It would have been too unkind of Fate to have it happen before.

No sooner did Charlie arrive than the whole earth seemed to erupt. A vast column of oil and gas soared skyward. The blast was so terrible that the derrick disappeared into the crater it made. The column of oil, 20 feet in diameter, gushed higher. Above the roar, Dry Hole Charlie could be heard yelling triumphantly, "My God, we've cut an artery down there!"

Drenched with oil, "Gusher" Charlie and his crew desperately tried to dam the flow, but there was no controlling it. The well gushed 125,000 barrels the first day and continued for months at the rate of 90,000 barrels a day. For 15 miles around, the countryside, buildings, and people were constantly sprayed with oil. A great dirt reservoir was built to catch the flow, while hundreds of men dammed canyon mouths to prevent flash floods from deluging the valley with oil.

Eighteen months later Lake View stopped as suddenly as it had started, after producing 9 million barrels. Dry Hole Charlie went back to drilling dry holes. Lyman Stewart, the explorer, was not a financier. He eventually lost control of Union Oil of California. But neither dry holes nor business manipulations could really matter to these two men. They had had their gusher.

4 : "A prospector is always rich"

CALIFORNIA'S PROSPECTORS streamed to the hills with the old-time abandon and enthusiasm of Gold Rush days after Lyman Stewart's first gusher flowed down the Santa Clara River. The constant talk of oil so stirred the imagination of Edward Doheny, a mining prospector, that when he saw an ice wagon loaded with tar jolting along a Los Angeles street, something prompted him to ask the driver about it. He learned that the tar had been dug from the Brea pits close by, and was being taken to the ice plant to burn as fuel. The driver directed him to the pits.

Doheny examined the black, sticky bogs with as much curiosity as the first Spaniards who had found the Indians using the tar to waterproof baskets, fasten arrow points to shafts, and glue decorative shells to ceremonial masks. With the descriptive simplicity of all discoverers the Spaniards called the area by a Spanish word for tar—*Brea.* Doheny knew that every adobe building in Los Angeles owed its roof to the pits. And yet, after centuries of use, the pits were as magically full as ever. As he mused over the point, insight replaced eyesight. The mining prospector follows outcrops in search of the mother lode. Doheny could visualize a mother oil pool feeding the pits, the oil thickening to tar at the surface. He sent an excited message to his partner,

Charles Canfield, who was laboriously tunneling into the rocks of their Southern California mining claim in search of gold. When Canfield arrived, Doheny presented his case like a lawyer.

"The mine can wait, Charlie," he argued. "This is new. We can dig here and find oil. That's what everybody wants now."

It was a curious partnership—the impetuous, daring Doheny and the gentle, patient, persevering Canfield. They matched each other in creative imagination and each was a born trail blazer, yet they achieved greatness only when working together. On his own, each had been a failure. Later, when death would dissolve their partnership, the spark of greatness in the survivor would die out in a scandal that would shock the whole nation.

Charles Canfield inherited venturesomeness from his father, a New York farmer who loaded his family in a prairie schooner to go to the California gold fields. Reluctantly realizing he was eight children too late for such a venture, the elder Canfield homesteaded in Minnesota and returned to farming. It was Charles, the fourth son, who, like an arrow from the parental bow, left the farm for the gold and silver mines of Colorado.

The twenty-two-year-old arrived just as the Caribou silver strike opened the silver seventies. When he made a profit of $6,000 on stock he bought in the mine in which he was working, young Canfield felt as rich as any of the silver kings. Returning home to share his good fortune with his family, he received an unexpected bonus. Falling in love with a neighboring farm girl, the confident prospector returned to the silver mines with a wife.

It was a bonanza time for everyone but Canfield. For fourteen years he found nothing but the opportunity to test and develop character. His wife and daughter had to spend more time on her parents' farm than with him. As he wandered from one unsuccessful mine to another, the theme of his poignant letters to his wife was that of every explorer who will not give up: "we are going to make some money here, *sure.*" •

How many other lonely failures did he speak for when in 1880 he wrote with pathetic cheerfulness from a mining shack in Nevada: "What did you have for Thanksgiving dinner? I got up an extra dinner today. I cooked up a chunk of turnip with potatoes—that's all the extra I had except what the mice dropped in the salt. How is the baby? I set her playthings up today where I could see them. Can she talk any more now?"

In his thoughts, he never failed. "Why, a prospector is always rich," he would say. "His strike is always just beyond the next break." When a fellow miner, with an equally long run of hard luck compounded by a broken leg, was starting off to look for that next strike, Canfield staked him with his last cent of cash. He was sure his friend needed the $6 more than he did.

Five years later he was writing his wife from another shack near Kingston, New Mexico. "Am still working with the usual results but with my old-time hopes. Now I can see how you look when you read this, but if I possess nothing more worth mentioning I have lots of faith and hope have I not?" This time his faith was to be justified. One spring morning, after months of toil, Canfield's pick broke through the next break into a rich vein of silver. Two stars had finally crossed. The claim he was working on for shares was owned by Edward Doheny.

Doheny had been as luckless a wanderer as Canfield. After graduating from high school in Wisconsin, he started his search for gold in North Dakota's Black Hills. Drifting from one mining camp to another, the insatiable need for the next grubstake prompted him to do other things from time to time. In Atchison, Kansas, he was the town surveyor. He traded horses in Oklahoma, soldiered in the Indian wars. Arriving in Kingston, New Mexico, broke as usual, he was delighted to fill the town's need for a schoolteacher.

The citizens soon realized the mild manners of this little blue-eyed man were deceptive, and that he could fill an even more urgent need as head of the vigilance committee. The frontier conflict between lawlessness and order was at its height. Doheny reveled in danger and excitement. He sent so many criminals to the penitentiary that the outlaws determined to eliminate him. He ignored their formal warning to leave Kingston. One afternoon a notorious cattle rustler, famed for his marksmanship, hid himself and opened fire on Doheny as he sauntered into range. Doheny ran, but not away. Paying no attention to the bullets, he dashed across the street and arrested his would-be murderer.

Canfield's strike on Doheny's claim seemed to be an end rather than a beginning of their partnership. Eager to make up to his family for their years of privation, Canfield sold his interest for $112,000 and moved to Los Angeles and luxury. Doheny, now that he had a prince's grubstake, was confident that he would become a silver king.

Putting all his money into real estate, Canfield repeated the error Lyman Stewart had made with his mowing-machine factory. When a Los Angeles land boom collapsed in 1890, Canfield's fortune vanished. He had often been broke, but this time he was heavily in debt as well. However . . . a prospector is always rich. At forty-two, looking for his next strike, Canfield shouldered his pick and shovel and returned to the Mojave Desert, staked by a $500 loan from an old friend.

He had already found his prospect when his star crossed Doheny's again. By this time, Doheny had gone through his silver stake and was looking for that next break, too. The partnership resumed.

When Canfield agreed to let the new gold prospect wait in order to dig for oil in the Los Angeles tar pits, the partners, unable to lease the land, spent their last dollar buying a corner lot. Neither of them had ever seen an oil well, so they cheerfully set to work sinking a shaft, 4 feet by 6 feet, mining for oil with pick and shovel as though they were after gold or silver. There was a new zest to handling the old picks, for the partners were intrigued by the novelty of their project. So were the astonished passers-by who watched them disappear into the earth.

When they had dug down 165 feet, their picks opened a layer of rock that started to flow 7 barrels a day. The partners rejoiced, little dreaming that their productive rock was only the top layer of a field that, like a monster wedding cake, lay 13,000 feet deep, with layer after layer waiting to flow more than 70 million barrels.

The shallow oil "mine" caused even wilder excitement than Lyman Stewart's Ventura County gusher of a few months before. Here was a way anybody could get rich, and everybody wanted to try. There was a stampede to buy and lease lots. Canfield and Doheny led the herd. Los Angeles was just settling down after the insane 1887 land boom. A railroad rate war, during which a ticket from Kansas City to Los Angeles cost $1, had done as much for California as the discovery of gold. Los Angeles' population jumped from 9,000 to 100,000.

The real estate circus that now started made the earlier one seem like a side show. Twenty-five-foot lots sold for $1,500 and West Los Angeles began to look like a gopher city. With oil selling at $2 a barrel, Canfield and Doheny were able to buy a drilling rig. Others used spring poles to kick down holes. There were so many frantic diggers that the dismayed City Council declared oil wells a nuisance and outlawed drilling in the city limits. Each operator promptly announced that he was drilling for water, a right which could not be denied. Within a few years 3,000 "water wells" were producing oil, and the Los Angeles City field was the state's biggest producer. The wells were so close together that there was almost more truth than jest in the story that a string of tools lost in one hole had been fished out of the hole next to it. There was so much oil, and still so little use for it, that the price plummeted to 25 cents a barrel.

Immediately after their discovery, Canfield left the new venture in Doheny's hands and returned to the mine. A prospector is always rich—that is, if he is always working. The Ophir mine lived up to its name. Once again, Canfield knew the heady thrill of discovery when his pick broke into a vein of pure gold. Doheny scarcely shared his excitement. He could think of nothing but oil. Canfield could not divorce so easily the companion of a lifetime to pursue a new siren. He suggested that he trade his interest in the oil field for Doheny's interest in the gold mine. Doheny was sorry to lose his partner, and thought Canfield was making a mistake.

He was right. The mine wasn't as rich as it had seemed. Canfield was soon back in Los Angeles, drilling two wells with Doheny with mediocre results. Always determined to turn defeat into victory, Canfield set out alone again into the hills, but this time in search of oil. After a string of disappointments he arrived in the Coalinga foothills of the San Joaquin Valley, 120 miles northwest of Los Angeles. Here he drilled some shallow wells that flowed 3 or 4 barrels a day. Prospectors catching fish this size usually moved on in disgust. Canfield decided to go deeper, on the theory that there was a "mother pool." A weary gold rusher had gone on record in 1850 that he wouldn't care to trade his old sweat-soaked saddle blanket for the whole west side of the San Joaquin Valley, thereby making it the most valuable saddle blanket in history. There were more than 6 billion barrels of oil there waiting for somebody like Charles Canfield to find them. His deeper well, which flowed 300 barrels a day, opened the first of the prolific fields.

Doheny, continuing to prospect, had found a fascinating guidebook to all of California's oil seeps. Five years before Lyman Stewart's 1892 gusher made California oil conscious, W. A. Goodyear, an assistant state geologist, had searched out almost every oil seep in the state, describing them in an annual report. Doheny marveled that it had not occurred to Goodyear to do anything more than describe them. "It can scarcely be credited that he knew the significance of the information which his annual report conveyed to the experienced prospector," Doheny wrote later. "That report was really my best guide in the discovery of the various oil districts which it was my good fortune to open up." The report led him to the Brea Canyon, some 20 miles southeast of Los Angeles, near Anaheim, which a German colony had made the best wine-producing area in California. For Doheny, the mountains "dripped sweet wine"—for the Brea-Olinda field was another giant, eventually yielding more than 300 million barrels.

Now that Canfield had been weaned away from mining, Doheny suggested they follow his guidebook together. The partners had the oil touch now. Wherever they drilled they found it.

No one knew, of course, the true magnitude of their discoveries at the time. It was enough for Canfield and Doheny and California that by 1900 the state's production had increased eleven times. Three years later the development of the Canfield-Doheny fields made California the nation's biggest oil producer.

It is not strange that this spectacular growth went almost unnoted by the rest of the nation. Big as their 1900 production seemed to Californians, it was only about 7 per cent of what the Oil Region—Pennsylvania, Ohio, West Virginia, Indiana, and New York—produced. In volume this was no competition, and the Rocky Mountains were an effective barrier to any commerce in California oil. Furthermore, California's product was thick and heavy, suitable principally for fuel; it yielded little of the only element in great demand at the time—kerosene.

These were the reasons why the outside world did not come streaming to California when Stewart, Canfield, and Doheny proved there was an abundance of oil there. These and one other. The pick-and-shovel digging in Los Angeles was a unique occurrence, as Lyman Stewart and hundreds of others who spent millions of dollars drilling dry holes could testify from frustrating experience. Only the most strenuous souls were attracted to such a risky business.

Stewart, Canfield, and Doheny were as ingenious in finding uses for oil as in discovering it. Since California was poor in coal resources, oil was welcomed as fuel by sugar refineries and small industries. Gradually, the cities began using asphalt residue from refineries to pave streets. Stewart went after the marine market. At first, steamships burning oil exploded, and its use was outlawed. After Stewart developed a safe burner, conversion was general.

Canfield persuaded the Santa Fe Railroad to experiment with oil from his Coalinga field. At the roundhouse, while mechanics were working on a special engine, a thoughtless workman inspected the oil tank with a burning lantern in his hand. The explosion destroyed the roundhouse, all the engines in it, and Canfield's hopes. Stewart experimented with engine oil burners and finally demonstrated that they could be used successfully. Canfield and Doheny were to benefit by this new use in an unexpected way.

The Santa Fe Railroad was so pleased with the economy of using the new fuel that Doheny shrewdly got the company to sign a contract guaranteeing to buy, at $1 a barrel, all the oil he and Canfield could find. The railroad was soon aghast at the partners' finding ability. Faced with the prospect of floating to defeat on a sea of oil, they offered to settle the contract by buying some of the partners' properties. Canfield and Doheny agreed to sell their new Midway field together with some unexplored leases. No greater tribute was ever paid to the belief in the Canfield-Doheny magic of picking oil territory than by the Santa Fe geologist. At a meeting to conclude the sale, he solemnly reported that the untested lands were capable of producing 250,900,000 barrels of oil. Canfield turned to Doheny and said with equal solemnity, "Ed, the son of a gun missed it by ten barrels."

5 : *"Something hidden. Go and find it"*

THE ASSOCIATION with Santa Fe opened the door to Canfield's and Doheny's greatest opportunity when A. A. Robinson, a Santa Fe official, became president of the Mexican Central Railroad. Knowing from experience what an asset it was for a railroad to have fuel available along its lines, Robinson invited the partners to Mexico in 1900 to see what they thought of prospects there.

As their private railroad car rolled along the central mesa of Mexico, the trip seemed more like a luxurious houseparty than an exploring expedition. Then the railroad plunged from the mesa down to the coastal rain forests, dropping by hairpin curves 5,000 feet in 30 miles. The vacation mood ended. The partners looked about with prospectors' eyes, not liking what they saw. Magnificent waterfalls and exotic jungle growth were not inspiring scenery; they were obstacles.

"This country is too difficult to prospect for gold, much less oil," Doheny commented to Canfield. "We're too far from supplies and too far from any market that needs oil."

"We'd have to find vast quantities to justify building pipelines, refineries, and tankers," Canfield agreed. "It would take so much oil, I doubt if we could sell it."

Their misgivings grew stronger when they left the comfortable private car and started through the humid jungle on burro back, looking for the oil signs their guide had promised them.

The bones of a Boston ship captain buried on the Mexican coast symbolized dramatically the failures of others who had tried to develop oil commercially here. For centuries the Aztecs had found many uses for the thick pure asphalt that oozed to the surface along the coast. In 1876, the captain bought some at Tuxpam for use on board his ship. Impressed with its possibilities, he persuaded some Boston friends to form an exploration company. A few holes drilled at great cost to about 500 feet found a little oil. The captain built a primitive plant to extract kerosene. This he sold to the natives. Returns were so pitifully small that the Boston investors refused his pleas for more capital. Discouraged, the captain committed suicide.

A few years later a British company, backed by Cecil Rhodes, abandoned another search after spending half a million dollars. They sold their holdings to a second British group, which spent another half a million just as futilely. Sir Boverton Redwood, one of the world's most eminent geologists, then wrote an obituary for Mexico's oil, concluding that it would never be found in quantities.

As their guide chopped a trail for them with his machete, Canfield and Doheny were overwhelmingly aware of the physical, economic, and historical facts against them. Then, suddenly, they saw a little spring bubbling oil on a small hill, and, as Doheny wrote later:

> The sight caused us to forget all about the dreaded climate, its hot humid atmosphere, its apparently incessant rains, those jungle pests the pinolillas and garrapatas (wood ticks), the dense forest jungle which seems to grow up as fast as cut down, its great distance from any center that we could call civilization and still greater distance from a source of supplies of oil well materials—all were forgotten in the joy of discovery with which we contemplated this little hill from whose base flowed oil in various directions. We felt that we knew and we did know that we were in an oil region which would produce in unlimited quantities that for which the world had the greatest need—oil fuel.

Doheny spoke for all oil explorers. The greater the gamble, the more it inflames the imagination and the more helpless the explorer is to resist it. He knows the rewards for success are in proportion to the risk involved. Though the desire for gain has always been a basic motive for all exploration, something more hounds the great adventurers. Like Kipling's explorer, they hear "one everlasting Whisper day and night repeated—Something hidden. Go and find it. Go and look behind the Ranges. Something lost behind the Ranges. Lost and waiting for you. Go!"

It was this Whisper that had always driven Charles Canfield and Edward Doheny. Now it seemed more like a shout when they learned from the guide that the bubbling oil spring prompted the first Spaniard who saw it to name the spot Cerro de la Pez—Tar Hill.

They were reminded of that other Spanish word for tar, *Brea*, which had been the beginning of the long journey that had finally led them here. As they studied the map, their excitement grew. From the Rio Grande River on the north to the Guatemalan border on the south, the Mexican coast was sprinkled with towns and hills whose names all meant tar or pitch—El Chapopote, El Chapopotal, Chapopotilla, Ojo de Brea. The early map makers had done almost as thorough a job in preparing a guidebook for the partners' experienced eyes as had the California state geologist with his survey of oil seeps.

Who had thought this country too difficult to prospect? With map in hand, with guides and pack burros, the partners spent 1900 searching out every oil seep they could find. Fifteen years earlier, Porfirio Díaz, Mexico's dictator, hoping to lure American mining capital, had repudiated the Spanish custom of reserving the subsoil riches to the government.

Mineral rights now belonged to the owners of the surface property. Whenever Canfield and Doheny, intoxicated by their high adventure, found a promising seep, they would buy all the land around it. At the end of their glorious spree they were the owners of 450,000 acres and a dream.

These two were kings among prospectors. With imperial indifference to all the incredible odds against them, they were impatient to risk on a new battlefield what it had taken them so many years to win elsewhere. In May, 1901, they stood again on Cerro de la Pez, contemplating the same little bubbling spring and listening to the sound of the tools pounding down their first Mexican hole. It was an expensive sound —they had spent $1 million in preparations. In Texas, 700 miles north of them, the great Spindletop boom was on. No one except Canfield and Doheny had time to be interested in what was happening in this small clearing hacked out of the lush Mexican jungle.

Two weeks after the well started, the driller awakened the partners at dawn, complaining, "There's so much oil coming into the hole, it's lifting the tools off the bottom and I can't drill ahead." He had been waiting for daylight to spring his little joke.

They had drilled only 545 feet. The cost, considering all they had spent, could be reckoned at $20,000 a foot. The well was not a gusher, flowing not much more than the 7 barrels a day that had filled their shallow pit in Los Angeles. Nevertheless, Canfield and Doheny were as joyful as Captain Lucas watching the mighty flow of Spindletop. They had won. Now, they told each other elatedly, it would just be a question of time.

Three years later they were wondering just how much time. With painful slowness they had drilled numerous shal-

low wells, none of them gushers, and produced no more than
125,000 barrels of heavy asphaltic oil. No one in America
shared their excitement or their confidence in the meaning of their discovery. One reason for this was that Spindletop had flooded the American market. The Mexican
Central Railroad, unimpressed by the amount of oil they
had found, politely refused to sign a contract. The two
gamblers continued to shove their chips on the table. If
no one wanted to share the risk, then their winnings would
be all the greater—perhaps. As fast as their wells in California flowed money out of the ground, they spent it in
Mexico. Three million dollars disappeared with astonishing
speed.

The railroad's refusal to buy their oil inspired the partners
to organize a paving company in Mexico City. Their heavy
asphalt crude was admirably suited for street covering and,
eager to create a market, they practically gave it away.
Mexico City and a half dozen other Mexican cities soon had
the distinction of being the best-paved municipalities in the
world.

Meantime, studying their maps to pick the location for
their next well, the partners assured each other again that
success was just a question of time.

6 : *"Sight along that arrow"*

TIME SEEMED to have run out for Michael L. Benedum, as
he dejectedly cleaned out his desk drawers in the First Citizens Bank of Cameron, West Virginia, one morning early in
1904. Alarmed depositors were clustered around the closed

front door, discussing the notice posted on it. No one had read it with greater bitterness than Benedum, the bank's president. At thirty-five he had just lost a fortune and now he owed one.

He went upstairs to the office of the Benedum-Trees Oil Company, to sort through the financial debris there. "Mike," said his tall, sturdy, red-headed partner, Joe Trees, "you were a fool to gamble with such dangerous things as banking and manufacturing. You should have put all your money in a safe business like oil."

Benedum could only agree. Like Lyman Stewart and Charles Canfield he was paying the price of being unfaithful to Lady Luck.

Benedum had started work at sixteen in a West Virginia flour mill after the depression of the eighties prevented his farmer-storekeeper father from sending him to West Point. Within two years he was a mill manager, but, seeking a field with more future, he started for Parkersburg in 1890 to apply to the railroads. On the crowded train, an act of courtesy changed his destination and his life. He insisted that an older man take his seat. The man, John Worthington, was head of South Penn Oil Company, a Standard Oil subsidiary. Impressed with young Benedum, he offered him a job. Mike Benedum, who had never heard of the oil business, asked if it had a future. Worthington smilingly assured him it had.

The South Penn company, one of Standard's first ventures into oil producing, was expanding. Benedum was sent out to obtain leases on promising land. His friendly, outgoing personality, his farm background, and his obvious integrity enabled him to lease where others failed. He loved the business, and delighted in learning about exploration and drilling. In four years he was Assistant General Land Agent in charge of all the company's activities in several counties in West Virginia and Pennsylvania.

When a new discovery is made, there is always a race to acquire all the open leases nearby. In one such race near Cameron, West Virginia, Benedum congratulated himself on having leased 1,500 acres, although the farmers, knowing how good the discovery was, insisted on payment of $5 an acre, $4 more than usual. Worthington was on vacation when Benedum advised the temporary manager of his purchase. The official demanded to know how much of the money was going into Benedum's pocket. Shocked and furious, Benedum blazed, "If this is the opinion you have, I think it's time for you to get another man." The official agreed with equal anger, telling him to take over personally all the leases he had made. "I'll do that, too," Benedum replied, and later refused to rejoin the company, refusing even Worthington's pleas.

Although he had to borrow to make up the $7,500 commitment, Benedum would soon be grateful to the only man who ever doubted his integrity. Benedum, his brother Charles, and a former South Penn engineer, Joseph C. Trees, began buying and selling leases and royalties. There was always a brisk trade in royalties since it was a means of speculating in oil without the expensive risk of drilling. When a landowner leases his oil rights, the usual arrangement is to reserve one-eighth of the oil for his share, or royalty, the custom dating back to when kings granted rights to exploit royal lands, reserving a share of the discoveries for the crown. Landowners are usually interested in selling part of their royalty as a hedge against the possibility of a dry hole.

South Penn employees were permitted to deal in royalties,

and such a deal had originally brought Benedum and Trees together. Trees had found a farmer willing to sell half the royalty on his farm, which was close to an oil discovery. He wanted $2,000. Tree's entire capital was $45. He discussed the situation with Benedum, who then put up the money for a three-quarter interest. That particular royalty had a remarkable history. There was oil on the farm and Benedum had received his investment back several times over when he sold it to his brother for a profit. After his brother made a good profit, he sold it to Trees. It returned *his* investment several times before he sold it back to Benedum. Benedum recovered his second purchase price and the royalty continued to pay on a small scale for thirty years more.

Six foot three, weighing two hundred pounds, Trees had Benedum's admiration because "he wasn't afraid of anything that walked, talked, or crawled" and had the same fearlessness in regard to working for success as an individual rather than being satisfied to work for a company. They complemented each other as partners just as Canfield and Doheny did.

Trees had stopped teaching school to work in the oil fields after he learned the least-skilled drilling hand earned a schoolteacher's monthly salary in ten days. Deciding that oil would be his career, he went back to college for an engineering degree. He worked his way through the University of Pittsburgh playing tackle on a professional football team for $75 a game. The athletic rules of the day did not prevent his also playing on the regular collegiate squad. He was an outstanding player. On one occasion when a professional commitment fell on the same day Pitt was scheduled to meet Penn State, the two universities agreed to postpone the collegiate contest so that Trees could play both games.

Buying and selling leases and royalties was a good beginning, but the new partners were looking for a lease that
warranted the risk of drilling. Trees finally bought a lease
for $400 in West Virginia and assured Benedum they could
turn it, undrilled, for $500. However, Benedum decided that
this was the one they had been looking for.

When word arrived that their first well was about to drill
into the sand where they expected to find oil, the partners
hurried to the lease. They paced the rig floor for hours with
emotions akin to those of expectant fathers in a hospital
waiting room. When oil began to flow from the hole they
couldn't even speak. They stood, grinning, scooping oil up
in their hands, and letting it trickle through their fingers.

"No man who has never had the experience can understand the feeling of exhilaration that comes when you
bring in an oil well, especially your first one," Benedum said
later. "There are no words to describe it. This may sound
strange, but the thought of the material profit to be derived
never crosses your mind. You are staggered and filled with
awe at the realization that you have triumphed over a stubborn and unyielding Nature, forcing her to give up some of
her treasure."

The partners brought in nine more producers, but their
triumph was soon tempered by as many dry holes on other
leases where Nature proved more stubborn. Meanwhile,
South Penn had found oil in the Cameron area where Benedum still held the leases that had caused his resignation.
Benedum refused to sell and he and the South Penn official
clashed again in a strange duel in which the weapons were
drills.

Benedum's scattered leases were in almost every instance
surrounded by South Penn's. In an effort to break Benedum

financially, the official ordered wells drilled where they would drain oil from under Benedum's leases unless he drilled offsetting wells. For every well Benedum started, South Penn started two more. The battle became an oil-field *cause célèbre*. Few men ever stood up so courageously against Standard as Benedum. He ran out of cash but not supporters. Drilling contractors, enjoying the fight, drilled for him on credit. Then he found banks to back him. He had nineteen wells drilling when South Penn finally realized he was their match. The battle ended in Standard's New York office when the twenty-seven-year-old independent was handed a check for $400,000 in payment for the leases he had bought for South Penn for $7,500 less than a year before. Again there was an exhilaration that had nothing to do with money-making. Mike Benedum had fought and won.

He and Trees formalized their loose partnership arrangement into the Benedum-Trees Oil Company with the understanding that each would always be free to act independently. Finding oil is so dependent upon men's freedom to disagree as to where it may be, that even partners must have this freedom. The lifelong partnership and successes of Benedum and Trees would prove the point many times.

Benedum's brother, Charlie, did not believe oil success was anything permanent and persuaded Mike to put a big part of his new fortune in pottery and glass manufacturing and stock in the First Citizen's Bank of Cameron. Mike became the youngest bank president in West Virginia—and quickly the most unfortunate one. The glass and pottery companies failed. Benedum endorsed their notes. The bank made loans that went into default. A young bank employee turned embezzler. Benedum-Trees drilled a series of dry holes.

As Benedum sat behind the bank doors that had failed to open, he was not thinking of the fortune he had lost, or the several hundred thousand dollars he owed. He was thinking about how to start again to explore for oil.

What is it that gives a Mike Benedum, a Charles Canfield, a Lyman Stewart the courage to make a fresh start when the fortune he has won is suddenly wiped out? Like a farmer who plows his field and plants it again when his crop has failed, they demonstrate a simplicity of faith and confidence in a natural law of supply, a profound belief that there is always abundance for those willing to work creatively. Dollars may disappear but the law of supply is always there to be used.

Mike Benedum was so confident of his ability that he refused to take bankruptcy, assuming responsibility for the bank's debts. His confidence was all the collateral he had, but it was good for a bank loan with which to start leasing again.

Having bought a lease adjoining a South Penn discovery well, Benedum sent for Joe Trees and the partners hired a horse and buggy to drive out to the land that would put them back in business. They were looking at a high stone ledge on which an arrow was carved, when a farmer came up to ask what they were doing. They explained that they were trying to pick a spot to drill for oil. The old farmer shifted the tobacco cud in his cheek and spat speculatively toward the arrow. "See that? Well, there's an old story they've told around here for years. Some pirates or robbers buried some treasure. They cut that arrow there to point out where it was. People have been digging to beat the band to try to find the stuff."

Mike turned to Joe. "Sight along that arrow. We'll put the

well right where it points." Joe agreed. The partners were
remembering a hunch they had played two years before.
Trees had stopped at a farmhouse for lunch, and the farmer
asked why he didn't lease the farm. Trees told him there
wasn't any oil for miles around. The farmer, who was blind,
asked to feel Tree's pulse. "You're all right, young man," he
said. "I'm going to tell you something I've never told anyone else. There's oil under this farm. Through my sightless
eyes I have seen it spouting over the top of the big maple
tree back of the house." Precognition? An old man's imagination sparked by desire? When the partners drilled, 300
barrels of oil a day spouted over the top of the big maple
tree.

This farmer was right, too. The biggest treasure the partners ever found in West Virginia was buried at the spot to
which the arrow pointed. Their well roared in for 3,000
barrels a day.

The year 1904 was as eventful on the national scene as it
was to Mike Benedum personally. This was the year Americans suddenly realized that the change of centuries meant
more than just the change of a calendar date. It was like
feeling spring in the air and then one morning awakening to
see the flowers and trees radiantly in bloom. There was prosperity everywhere. No longer were the luxuries of telephones,
electric lights, and plumbing exclusively for the very rich.
Industries were learning the art of mass production with
cheaper costs. Popular magazines, their circulations leaping,
began advertising a bewildering abundance of improved
necessities and fascinating new conveniences.

The year before, Orville and Wilbur Wright had made
man's first powered flight. The historic event caused little

general excitement. Everybody was automobile mad. There were 178 automobile factories employing 12,000 workers. Two pioneers drove a Franklin from San Francisco to New York in 43 days. John Wanamaker's department stores in Philadelphia and New York sold automobiles, in addition to all the appropriate clothing to wear in them.

"Everyone can afford a Fordmobile," Henry Ford proclaimed. In June, 1903, eleven years after making his first car, he had started again at forty with his third company, and by March, 1904, had sold 658 Model A's at $750 apiece. The car was "built for business or for pleasure," his advertisements stated. "Built also for the good of your health—to carry you 'jarlessly' over any kind of half decent roads, to refresh your brain with luxury of much 'outdoorness' and fill your lungs with the 'tonic of tonics'—the right kind of atmosphere."

The lungs were perhaps more favored than the rest of the body, for the first census of American roads in 1904 revealed that you could be carried "jarlessly" over only a few of the country's 2 million miles of roads. Ninety-seven per cent of the mileage was dirt, and what were called "improved highways" were mainly gravel.

Though there was little doubt that eventually all America would be on wheels, nobody—least of all the oil industry—anticipated how quickly the automobile, this child of the new century, would come of age. It would be another five years before Henry Ford would announce his Model T, the car that in most history books would take precedence over the invention of the automobile as the date of the industry's birth.

John D. Rockefeller and his Standard Oil associates, now

controllers of 82 per cent of the nation's oil business, were actually worrying about markets for their products in the midst of all this glorious change and prosperity. Gasoline was being thrown away at refineries as an undesirable product. Who could foresee that the spluttering little buggies that could drive up to 50 miles on a gallon of fuel would soon number in the millions, creating a greater market for gasoline than all the lamps and stoves of America had ever created for kerosene? Or that almost 2 million miles of dirt roads would need to be paved with oil to support their incessant travels?

Like sailors caught in a storm, the Standard officials were too busy with their sea of troubles to think about their future course. What they had done to the whale, electricity was now doing to them. The Kerosene Age was ending. The new Fuel Age was a decidedly mixed blessing from their viewpoint, for it wasn't Standard oil that was fueling and lubricating all the new factories, trains, and ships. Captain Lucas, Canfield, Doheny, and Stewart had seen to that. The vast quantities of oil these explorers had discovered in Texas and California had doubled the country's oil production between 1900 and 1904, and Standard did not control the increase. Further, California and Texas crude oil was almost pure fuel oil from the well, producing little kerosene.

To make matters worse, Theodore Roosevelt, just elected to his second term, was waving his big stick more threateningly than ever at Standard. In 1903 Ida M. Tarbell's blisteringly critical *History of the Standard Oil Company* ran serially in *McClure's* magazine, crystallizing public opinion against the monopoly. The resulting opposition encouraged competition. Oil was now a game any gambler could play—

the machine was no longer fixed in favor of Standard. There
was a whole country in which to look for oil, not just a tightly
controlled Oil Region in the East.

Mike Benedum and Joe Trees were ready. Indian Territory, soon to become Oklahoma, looked promising. In July,
1903, two diggers in the territory struck a 50-barrel well.
Like schoolboys racing after a fire truck, Benedum and Trees
hastened to join the rush of oil seekers. The blaze proved
more of a grass fire than a conflagration. The partners drilled
a well and found a little oil, but lack of a market and government restrictions on leasing Indian lands cooled their enthusiasm. They were happy to sell at a substantial profit.

When an old friend of the partners walked into their West
Virginia office and recommended going to Illinois, he might
just as well have suggested the Belgian Congo. There was
no production in Illinois and the friend had more enthusiasm
than evidence of its prospects. He had drilled one hole that
was producing gas and a little oil. The idea of gas intrigued
the partners. There would be a market for it, they reasoned,
in such a populous area. But the head of a Pittsburgh gas
company who went with them to inspect the friend's well
told them flatly it was no good.

Nosing around the countryside, the partners picked up
the trail of John Worthington, the South Penn manager who
had started Benedum in oil. He gave Benedum a friendly
warning: "Don't get burned, my son. We've gone through
this region pretty thoroughly, and are convinced that it's no
good." The partners were inclined to take his advice, but
still they wanted to tramp the creek beds themselves and
study the rock layers through which streams had cut.

Geology was a new science in 1904 as far as practical oilmen were concerned. A popular saying was, "Geology has

never filled an oil tank." So far, of course, this was true. Early oil explorers followed the ancient mining principle of looking for larger quantities of a mineral near where it had been found on the surface. Lacking surface indications, geologists usually stated where they thought oil couldn't possibly be, rather than where it might be—the method that had caused Captain Lucas such anguish at Spindletop. The explorers, rather than the geologists, had to supply the facts and the impetus to use them. Petroleum geology lay dormant as a science until the explorers, finding oil in great quantities, created such demand for even greater quantities that the science had to be developed to meet the need.

As Benedum and Trees examined rocks along Illinois streams, they were applying "creekology" and "trendology," the theories that practical oilmen had devised in the forty-five years of exploration since Drake drilled his well on Oil Creek. They were looking to see in what direction the rock layers slanted, indicating whether or not they would be tilted underground to form a trap for oil. They were looking to see if the rocks were porous sandstones that could be a reservoir for globules of oil, like a comb for honey. And, as the Pennsylvania and West Virginia oil traps were found in a trend running northeast to southwest, they were looking for this, too. They found indications that there were traps underground, but were puzzled that they lay in a northwest-southeast trend. They discussed the unorthodox idea that in different regions oil traps might lie in different directions.

Confirmation of their idea came from an unexpected source. There were perhaps only half a dozen geologists in America then concentrating on the study of oil. A hotel lobby in Casey, Illinois, was an unlikely spot to meet one of them, yet there he was, a resplendent fellow in Prince Albert coat and high silk hat. In the quick intimacy a small-town hotel lobby generates among strangers, the partners told him their idea. He earnestly agreed, explaining that he was writing a paper detailing his belief that this was oil country. To prove the trend was contrary to old Pennsylvania–West Virginia ideas, he said that in addition to token quantities of oil found near Casey, some moderately productive wells had been drilled 70 miles southeast and at Robinson, midway. If they would go to Robinson they would find a retired blind judge who had drilled a well with a show of oil and had long been trying to interest someone in more drilling. The geologist had examined samples from all these wells and found they were the same rocks. It should take only the right location and perhaps deeper drilling to find a great oil field. Benedum and Trees were so enthralled with the picture the geologist painted that they forgot his name. Unfortunately, they never saw him again. It was a curious encounter between geology and practical oil finding. Its extraordinary results would be prophetic of the permanent union to come.

Remembering another sightless man who had foreseen oil gushing over a maple tree, the partners went to Robinson as fast as team and buggy could take them. The blind judge substantiated his vision with maps, well samples, and detailed history of the local efforts that had failed owing to inexperience and lack of financing.

Experiences of great explorers, when looked back on, usually seem to have been pitched on an eventful plane of high excitement different from the "life is so daily" course of other men. Actually, to the explorer, the unfoldment generally seems slow, proceeding from the most ordinary things. Captain Lucas reading an ad in a newspaper. Doheny idly watching an ice wagon jolting down a street. Like the building of a coral reef, each new venture grows logically out of and on the one just behind. Yet to each explorer at some time comes a great moment of decision, an exhilarating awareness that he must prove himself worthy of his calling. This is the great gamble. If taken, it must be taken premeditatedly, and with confidence in the taker's ability to survive failure. Captain Lucas knew this moment after his first well failed at Spindletop. A little spring bubbling oil in a Mexican jungle was the challenge to Canfield and Doheny. Benedum and Trees now faced their moment in Illinois.

They knew Standard experts considered the area worthless. They had seen for themselves that the small amounts of oil found were noncommercial. They knew their theory was contrary to everything known and proved about where oil might be. And even in the proven area of West Virginia they had drilled many dry holes. A few months before they had been penniless. Now they had a comfortable fortune, thanks to their well on the West Virginia farm and the sale of their Oklahoma property. They could continue to prosper by trading leases and drilling likely ones in an area they knew well. To drill in Illinois was not a question of just one well, which they could easily afford to forget if it were a dry hole. To warrant drilling even one, they knew they must be prepared to drill many. Why risk security against such odds? No reason—except that they were explorers.

Columbus did not attend to the details of outfitting his ships with any more rising sense of excitement than Benedum and Trees prepared their campaign. Since they were going to risk all their capital and persuade others to join them, they decided to lease as much land as possible before drilling. Again they departed from custom. Instead of leasing a few hundred acres, they leased 50,000. "You can get just as dry a hole on forty acres as you can on four thousand," Benedum maintained, "but when you get one on four thousand you've got a lot more chance for making a strike later."

The first well, in a ravine near Robinson, was a dry hole. This obviously wasn't going to be any easier than they had thought. The eager "creekologists" followed the stream bed to a spot where rock formations indicated they were higher up on the underground trap they hoped was there. This time, on August 5, 1905, from a depth of 950 feet oil shot into the air: 25 barrels a day, with 1,500,000 cubic feet of gas roaring up with it. To Benedum it was "the prettiest sight I had ever seen in my life." The next well, which gushed 2,000 barrels a day, was an anticlimax by comparison.

Illinois was the biggest oil news since Spindletop. The usual swarm of boom-town vultures settled over the fresh kill, but their predecessors at "Swindletop" had educated the public. This time there was more oil-finding than stock-printing. Actually, there was so much oil everywhere, such ventures might have been profitable. Whereas Spindletop was confined to a few acres, the Illinois fields spread in every direction. Within two years the state was producing twice as much oil as Texas and almost as much as all the states of the old Oil Region combined.

Standard's lease men were paying bonuses of $200 an acre for leases in the "no good" territory where farmers had gladly leased to Benedum and Trees for 10 cents an acre. They even bought some of the partners' producing properties for $350,000. The purchase illustrated the unpredictability of oil values. The property included a 2½-acre lease that Benedum's brother-in-law, Jim Lantz, had brought in apologetically during their leasing campaign, saying it probably wasn't worth the recording fee but the farmer was so desperately poor his sympathy had moved him to pay $25 for it. Benedum gave it to Lantz as a bonus for his work. When Standard, through its subsidiary, Ohio Oil Company, bought this particular group of leases, Benedum insisted that Lantz's lease be included for $25,000. Ohio paid reluctantly. Twenty years later the Standard official who handled the purchase told Benedum that the 2½-acre lease eventually produced enough oil to return the $350,000 paid for all the leases. The royalties also turned the impoverished farmer into a prosperous one.

The big gamble of Benedum and Trees proved that the indispensable tools in oil exploration are freedom to disagree as to where oil may be and courage to risk money to prove your belief. If the rewards are not commensurate with the gamble, courage to risk will be lacking in the same degree. The partners' rewards in Illinois matched their courage— they netted almost $8 million. Now was the time to quit if profit making was the only incentive. However, the game itself meant equally as much to them, and with their Illinois profits they could play for even higher stakes. They decided to risk their luck once more in Oklahoma.

7 : *Independence Tea Party*

Early in the century Kansas was symbolized in the minds of Americans by the gaunt form of Carry Nation, in a black alpaca dress and black poke bonnet, striding into saloons shouting "The arm of God smiteth!" as she smashed everything in sight with a hatchet. With the same spirit of raw pioneer wrathfulness, Kansas oil producers astounded everyone by attacking the great Standard Oil Trust. In March, 1905, they held an indignation meeting at Independence and their "new Declaration of Independence" marked the turning point in public opinion, the force of which eventually smashed the Trust as effectively as the woman with a hatchet did a mahogany bar.

Kansans were hopping mad. "This is not a war on the Standard, but a war against its methods and business policies," Governor E. W. Hoch thundered ambiguously. "In this contest we will put every dollar of our resources, every bit of our manhood and womanhood and every particle of governmental power and we will fight it out until victory shall finally crown our efforts. The eyes of the civilized world are on Kansas and I have been flooded with letters and telegrams of congratulations. The heart of the Union is with us and President Roosevelt is with us heart and soul." As Governor Hoch was an oil producer, he also was with the cause heart and soul.

"When the vulture leaves the carcass, the Standard will leave Kansas!" cried a producer. Delighted reporters for New York, Boston, Philadelphia, and Chicago newspapers filed many thousands of words from the meeting. Not since the days of the Bryan-McKinley campaign had the country heard such fiery, intemperate oratory.

The fact that Kansas was producing enough oil to fight about was as surprising news as its self-appointed role of David to Standard's Goliath. There had been no sensational discoveries in the state. In 1893, W. M. Mills had walked into the Pittsburgh office of John Galey and James Guffey seeking their financial help to drill in Kansas, just as Captain Lucas was to seek their help for his Texas venture seven years later. Mills had a little bottle of oil, the magnet that could draw Galey and Guffey any place. However, Mills had no Kansas Spindletop. Galey and Guffey had the frustrating experience of finding some oil almost everywhere they drilled, just enough to keep them looking for something better. They drilled almost a hundred wells and spent half a million dollars. Having no market for the oil they could produce, they were relieved to sell their leases, equipment, and production to Standard for half the amount they had invested. Standard efficiently doubled the production, built a refinery, and developed markets so competently that by 1903 the Kansas City *Star* could report that "Kansas, struggling a few years ago to pay off a heart-breaking load of indebtedness, is alive today with the hum of factory wheels, the flare of furnaces in foundry and glass plants, brick yard and smelters, flour mills, sorghum mills and cotton mills."

This prosperity, achieved on a modest production of 2,600 barrels a day, stimulated such a hunt that the following year four hundred independent companies were producing 11,600 barrels a day. Standard, the only buyer, cut the price of crude in half, and refused large quantities. It preferred to buy oil of better quality from the new discoveries in Oklahoma. At the same time, railroads increased freight rates and Kansas oil could not be marketed competitively outside the state.

Kansas oilmen were in no mood to accept Standard's view of what had happened. "It is the old, old story with the Kansas producers, as well as with other producers," Daniel O'Day, a Standard veteran explained. "As they get too much oil, they must expect the price to decline as in the case of an over-crop of corn or cotton, or any other commodity. When the prices decline they naturally kick very hard."

Standard underestimated how hard Kansas could kick. A featured speaker at the Independence mass meeting was Ida

Tarbell, whose accusatory *History of the Standard Oil Company* was then the most discussed book in the nation. She had grown up in the Oil Region, where her father had been an unsuccessful producer. She spoke these men's language. "Stop sizzling," she exhorted them, "and play the game as well as Standard Oil plays it. Your problem is to get in touch with the world market. You cannot do this by cursing Standard Oil. Play the game with the energy with which Kansas men can play a game, but play it like gentlemen, that is with due regard for the rights of men, something the Standard has never done. Get down to business. The time you spend in talking in Independence, the Standard Oil Company spends in putting up one or more fifty-thousand-barrel tanks and laying ready for use ten or twenty miles of pipeline and refining tens of thousands of barrels of oil. It is keeping quiet and doing business."

The oilmen translated oratory into action. With the speed of a Kansas tornado they succeeded in getting the state legislature to pass laws establishing maximum freight rates on oil and prohibiting rebates and discrimination in purchase prices. Pipelines were made common carriers. In an excess of enthusiasm, the legislature even authorized a bond issue for a state-owned refinery using convict labor to compete with Standard. This socialistic move was later judged unconstitutional and, as events would prove, it was unnecessary.

Applauded by a nation whose social conscience was ripe for active revolt against monopoly's restrictions, the Independence Tea Party inspired federal investigations that were to bring about the legal dissolution of the Standard Oil Trust within six years.

The oil producers, however, scarcely bothered to think about what the federal machinery might accomplish. They considered the Trust already smashed, or at least badly beaten. Actually, in a way they did not fully realize, they had already accomplished a major part of their objective, regardless of the legislation to come. Abundance is the true archenemy of monopoly, and the Independence meeting was historic in the confidence it created among oil explorers. Oil is found in abundance only if a great many people are looking for it at once in all sorts of unlikely places. The confidence that they could sell oil in spite of Standard, and build new pipelines and refineries if necessary, gave fresh impetus to the oil search. From the Kansas battlefield emerged a rugged new breed of young independents who would spread out over the mid-continent regions looking for oil with the fervor of crusaders.

One of these young men was twenty-nine-year-old Harry Ford Sinclair, who had been a pharmacist in his father's Independence drugstore until the discovery of oil near the town showed him his career. He then bought and sold leases for an oil company until he accumulated enough capital to start exploring on his own. Of all the oilmen who listened to Miss Tarbell that notable day in 1905, Harry Sinclair was the one who would follow her advice to the letter, becoming one of Standard's greatest competitors in all its fields of operation. Three months after the Independence meeting, almost on the very day that Benedum and Trees were jubilantly viewing in Illinois "the prettiest sight" of their lives, Sinclair was experiencing the same emotion in Kansas—at the sight of his first producer.

Like Napoleon, whom he resembled in stature as well as character, Sinclair was a citizen born to be emperor, and he knew it. He was bold, coldly shrewd, and had such conviction of his own superiority that no man dared to dispute

it. He did not blunder into success. He acted directly, with
a furious energy that drove others as unsparingly as he drove
himself. These were not latent qualities, developed slowly
by experience, seasoned and tested by error or defeat. He did
not acquire power. He simply exercised it.

Harry Sinclair knew only one direction to go—up—and
that in a hurry. The time was ripe for such a man. When
he heard of a new strike on the Ida Glenn farm near Tulsa,
in Indian Territory, Sinclair went there so quickly that he
was able to buy some of the choicest leases before prices
rocketed beyond the reach of his budding new bank account.

A sleeping giant was waking in Oklahoma. The shallow
wells were flowing from 1,000 to 3,000 barrels a day, and the
product was causing a sensation because, unlike the heavy
fuel crudes of Spindletop and Kansas, the Glenn Pool oil
was rich in kerosene. Standard Oil immediately began laying a pipeline from Kansas. When a bottle of wine was
ceremonially broken over the last connection, Glenn Pool
oil flowed to the Atlantic Coast through the longest pipeline
network in the world. But even before the line's completion,
it was inadequate to handle the oil flood that Sinclair and
others were developing. In a little over a year the pool's
producing wells covered 8,000 acres. Help to the producers
was on the way in the form of two pipelines snaking up
from the Gulf Coast as fast as Gulf Oil and The Texas Company could lay them.

Glenn Pool, in 1905, was the salvation of the Mellons'
$5 million investment in Spindletop. Their wells in Texas
had practically ceased to flow. Even if they could have
bought all the oil being produced in that state, it wouldn't
have kept their big Port Arthur refinery going. Besides,
Texas oil was of such poor grade there was little use in

refining it. It could be used for fuel just as it came from the well. Gulf's pipelines and fleets of tankers were idle. But now, if the Mellons' could tap the new Oklahoma field 550 miles away, they could operate their Texas refinery in competition with Standard. They could ship products to the Atlantic Coast by tanker cheaper than Standard, which had to go overland all the way. Enthusiasm quickened when the Mellons discovered that Glenn Pool oil was potentially rich in gasoline. The Mellons were the first producers to appreciate the possibilities of gasoline. They had their eyes on the French market, where the automobile industry was already flourishing, and it seemed to them the same development was about to happen in America. Andrew Mellon's genius for knowing how and when to put money to work constructively brought its greatest results in his decision to invest $6 million more to stay in the business acquired "through the vicissitudes of banking." The Mellons were not interested in becoming another Standard Oil; they planned to be explorers as well. Their purchase of producing properties in Glenn Pool would get them started. The Mellons had a new vision of bigness in oil.

Harry Sinclair had it, too. Of course, Gulf was starting with a $15 million company while he was still working on his hardest million, the first. However, it would not be long before he would catch up with Gulf.

The oil emperor-to-be held unique court in the crowded lobby of the Tulsa Hotel. The little cow town, which had earlier prospered on false oil booms, was now surging with producers, prospectors, salesmen, and adventurers from Kansas, Texas, West Virginia, Pennsylvania, Indiana, and Illinois. The hotel lobby was the boisterous market place where men traded dreams. Sinclair held himself as aloof from the shoving, excited bargainers as though he were already John D. Rockefeller. He was commuting daily by train from Independence and had rooms at the hotel. Just before train time, the impecably dressed little man would stride imperiously through the sweaty, rough crowd of lease buyers and sellers, not bothering to look at them. Inevitably, an eager seller would dare to stop him to offer a 40-acre undrilled lease. Sinclair, who had his campaign map in his head, would pause briefly, snap, "Wouldn't pay you a dime over fifty dollars an acre, but you can come see me tomorrow," and sweep on out toward the depot. On the train to Independence he would chuckle at the knowledge of what was happening in the lobby. He knew that the going price of acreage in the area of the offered lease was only $25 an acre and that, by his offer, he had effectively tied up under free option every acre anyone had to sell. When he returned to Tulsa the next day there would be a crowd of brokers waiting to sell him everything available. He would be able to see if he could buy a large enough spread of acreage to be of interest. Meantime, he could evaluate the area's prospects. He was not committed to buy and, confident of his skill as negotiator, he knew he could bargain the price downward to whatever he decided to pay. This man knew what to do at every moment. Other men had no plan, and since they followed events rather than creating them, they found themselves following Sinclair.

Oil pools, big and small, were found in an ever-widening circle from the Glenn Pool discovery well. Drilling was cheap, for oil was found at shallow depths. Sinclair would drill a choice lease, prove it for oil, then sell it, often to Standard, to get money for bigger plays. Others were pulling themselves up by their bootstraps this way, too, but Harry Sinclair's were seven-league boots.

The spirited competition for leases was flavored with hilarious melodramas. A classic incident, typical of the times and of Sinclair, is related by W. J. Connelly in his *The Oil Business As I Saw It.* Connelly shared Sinclair's oil adventures for half a century. In the fall of 1906, explorers found oil in Indian Territory near the Kansas border. Leasing Indian lands was complicated by Department of the Interior restrictions. Frank Tanner, a Cherokee, owned a 110-acre allotment near the new production. Twenty-year-old Frank was a minor; the Department permitted leasing of his land, but the contract would terminate on February 14, 1908, the day Tanner came of age. He would then be free to lease on his own terms. The cautious lease owner was unwilling to risk drilling to prove the acreage that might then be sold to somebody else.

Sinclair had a hunch there was oil on Tanner's allotment. He instructed Connelly to offer Tanner a $20,000 bonus if he would agree to sign a new lease to Sinclair when he came of age; until then he would be paid $100 a month on account. Tanner signed the agreement. Sinclair, knowing it was not legally enforceable, was delighted to find a more subtle way to bind him. The young Indian was a promising baseball player, and Sinclair had bought the Independence Baseball Club. Tanner was signed as pitcher for the next season's team. Connelly, nursing the situation along, noticed signs of Tanner's becoming "nervous." The manager of the Pittsburgh Pirates in the National League, a friend of Connelly's, was persuaded to write Tanner a letter promising to give him a tryout the following spring. This kept Tanner happy.

The week before the crucial birthday Sinclair and Connelly also began to grow nervous. Sinclair did not have $20,000 on hand to pay the bonus. On a trip East he obtained a loan and wired Connelly to get the lease. Connelly went to Tanner's home at Nowata and was dismayed to find him gone. He caught sight of an enterprising lease broker carrying a grip and followed him on a train to Coffeyville, Kansas, thinking perhaps he had hidden Tanner. It was a false trail. Connelly went to Tanner's parents' farm, learning there that Tanner and a friend, Joe Rogers, had left on a trip. A hack driver who had driven them to the Katy depot that morning said they had bought tickets for Guthrie, 130 miles west. Connelly caught the next train, telling the conductor to wake him to change trains. Forgotten by the conductor, he woke up in Oklahoma City, 25 miles beyond Guthrie. Waiting for a train to go back, he called a friend in Oklahoma City and asked him to check all the hotels for Tanner.

In Guthrie, Connelly learned that Tanner and Rogers had checked out of the hotel the night before, saying they were going to Denver. "This was a stunner," Connelly relates. He knew only one man in Denver, an attorney, whom he called. The attorney promised he would have the Denver police pick up Tanner and hold him. At this point Connelly's Oklahoma City friend telephoned that he had located Tanner and Rogers in the company of Charles J. Wrightsman and Gene Blaize, "two well known and smart oil producers." Connelly knew there was trouble ahead.

Rushing to Oklahoma City, he found his friend had been jailed when Wrightsman, observing that he was being followed, complained to the police. Connelly and his friend learned later that the oil producers were staying in a hotel, the Indian boys in a rooming house. Connelly trailed the boys to the hotel the next morning, and when Rogers, apparently in the oil-producers' hire, went upstairs to confer with them, Connelly cornered Tanner in a washroom. Brushing aside his attempts to explain why he was in Oklahoma City, he told him the manager of the Pittsburgh Pirates wanted him to come to Pittsburgh at once. Assuring him he needn't worry about money, clothes, or telling his friend good-by, Connelly rushed him to the depot. Four minutes before the train pulled out, the enraged oil producers and the Indian friend arrived sensationally in a Thomas Flyer, one of the few automobiles in Oklahoma. They boarded the train, snatched Tanner, carried him to their car, and roared off. "This," Connelly complained, "was a body blow."

He telephoned the police chiefs in all the larger towns nearby asking them to locate the Thomas Flyer, "owing to an emergency affecting one of the men in it." Reluctantly, he telephoned Sinclair and gave him the bad news. "You can handle Blaize," Sinclair decided, "but Wrightsman is too fast for you. I'll go down to Oklahoma City and join you."

A bribe prompted a garage man to report the rented Thomas Flyer's first night stop. The town was between Enid and Guthrie. As the next train left for Guthrie, Sinclair and Connelly rushed there on the chance it might be the kidnapers' destination. Sinclair managed to borrow one of the few other cars in Oklahoma—it belonged to the new state's lieutenant governor—and made ready to pursue the Thomas Flyer. By nightfall, after much telephoning, Connelly located the fugitives in Enid, 73 miles away. Rain, mud, and darkness made Connelly and Sinclair give up the idea of driving. The last train had left, but the stationmaster agreed to run a special train. Not having the price of it, Sinclair and Connelly scouted the hotel lobby and found a Kansas friend with $25 cash and a draft for $200. The train was on the track ready to go, but the stationmaster refused to cash the draft. The hotel cashier didn't have enough money and they had to get the owner out of bed to make up the difference. Back at the station they found the stationmaster had dismissed the train crew. When the train finally pulled out it was almost midnight. Tanner's birthday was only minutes away.

In Enid, the police chief revealed that Wrightsman was at one hotel and Blaize, with the two Indian boys, at another. He agreed to pick up Tanner and have him in his office first thing in the morning. It was 2 A.M. when Sinclair and Connelly pounded on the door of Wrightsman's hotel room. "Wrightsman was utterly dumbfounded," gloated Connelly. " 'I thought I left you at the Katy station in Oklahoma City,' he said. I agreed this was where he had seen me last but added that I had moved around some."

The men argued futilely for several hours. Wrightsman agreed to meet Sinclair later for more parleying, promising to make no move meanwhile. At 8 A.M. the police chief telephoned the reassuring news that Tanner was in custody. When the rivals resumed their conference, Wrightsman simply laughed at Sinclair's proposal to settle the matter by each taking a half-interest in the lease. He was willing to let Sinclair have a quarter at most. Connelly then had the police chief produce Tanner, who went directly to Connelly, declaring "I want you to have my lease."

That afternoon, Sinclair and Wrightsman commemorated the chase by forming the Chaser Oil Company in which each had a half-interest. Tanner suffered the only disappointment. He never became a big-league pitcher. He could have bought a ball club if he wished, however, for his lease proved to have 2 million barrels of oil under it.

One day in the spring of 1909, Sinclair waited at the Independence station for a special train from Kansas City. He was as immaculate and impressive-looking as the Buick touring car he had bought for the occasion. A crowd had gathered around it. Any car was still a curiosity, but this was the most elegant horseless carriage Independence had ever seen.

Although he did not show it, Sinclair was excited. In a few minutes John D. Rockefeller, Jr., and a party of Standard Oil officials would be arriving. It was the thirty-five-year-old Rockefeller's first visit to the mid-continent oil fields; Sinclair was to be his guide on the party's brief stop. Sinclair had read the interview Rockefeller had given in Kansas City the day before, in which he excused himself from making any comments on oil matters by saying, "I don't know anything about the oil business although I am a director of the company. I look after my father's personal affairs and that is quite a little chore. I found that it kept me so busy that I had to give up my Bible class."

Rockefeller was much more voluble on the subject of a young man's chances in America, which he thought were "first rate" if he had "ability, integrity and is on the square." "There are so many young men who like to live on easy street who do not like to do a day's work and there isn't a chance for them any place. But there are not so many of that sort out here as we have in the East. They are a fine lot out here. I have found that out and from what I can find out the chances are better here and the young men are more willing to embrace the chances."

The thirty-three-year-old Sinclair knew all about the willingness to embrace chance and its results. What he was thinking about was how to convince this other young man that Standard should give the young men of Kansas even better chances by increasing the price of oil. Feeling against Standard was more violent than ever in Kansas. Sinclair, as treasurer of the Mid-Continent Oil and Gas Producers Association, intended to make good use of the time he would spend with young Rockefeller in the back seat of his new Buick showing him the town.

The Standard Oil party was affable and gracious. Rockefeller was gratifyingly interested in all he saw. Sinclair found an appropriate opening and plunged into a detailed description of the oil producers' plight. He thought Rockefeller was paying close attention, but apparently he had not been.

"Is this your automobile, Mr. Sinclair?"

"Oh, yes."

"It's a very beautiful one."

"Thank you, Mr. Rockefeller. It's the first Buick in Independence."

"Well, Mr. Sinclair, it seems to me that if you have been able to buy an automobile like this with the present price of oil, that you producers are doing very well."

Sinclair remained silent. Talk was a waste of time. Ida Tarbell had been right when she said the producers' only solution was to play the game as well as Standard Oil played it. "Not just as well," said Harry Sinclair to himself. "I'll show them how to play it better."

8 : Chaos underground

MIKE BENEDUM and Joe Trees were having a dry spell. After the sale of their Illinois properties they had decided to try Oklahoma again. The year was 1907, and Glenn Pool had made Oklahoma the nation's biggest oil producer. But Oklahoma just wasn't for them. This time it cost them $50,000 to find it out. They leased several thousand acres, but the rock formations posed such a difficult drilling problem they abandoned their exploratory well. Had they managed to go 130 feet deeper, they would have found a giant oil field. Its discovery five years later set off Oklahoma's greatest boom. This missed opportunity taught the partners a profitable lesson. Thereafter their policy was like that of another explorer who, failing to discover a pool by one foot, always drilled his wells one foot deeper than where he planned to stop.

The next move of Benedum and Trees was typical. Other explorers and companies, thinking Louisiana was finished as an oil state, were breaking camp and leaving for Oklahoma. Disappointed in Oklahoma, Benedum and Trees went to Louisiana.

For a short time after Spindletop's discovery, Louisiana had stolen the show. Scott Heywood, the ex-vaudeville actor who drilled Spindletop No. 2, the hill's greatest gusher, was approached by a group of businessmen from Jennings, Louisiana, 90 miles from Beaumont. They had leased 2,000 acres there, around an old gas spring, and they wanted Heywood to drill a well. Nine months after the great strike in Texas, Louisiana had its first gusher. Jennings was another salt dome, almost as rich as Spindletop.

The following year Ellison M. Adger drilled 425 feet in the Caddo Lake area near Shreveport. Looking for fresh water, he found salt water. Still hopeful about the area, he sent soil samples to A. C. Veatch of the United States Geological Survey, requesting advice. To his disgust, Veatch's answer was that there was no water where he had drilled, but if he would deepen his well to 1,000 feet he would undoubtedly find oil or gas. Adger thought this "sounded like a fairy tale." Besides, he wanted water, so he went someplace else.

Veatch, who had studied the area, redeemed the Survey's oil reputation which his chief had lost at Spindletop. When oil seekers drilled at Caddo in 1905 they found a new phenomenon—wild gas wells that blew out, leaving great craters. There was oil, too, but gas roared burning from the wells in such enormous quantities that at night the countryside was theatrically lit for 25 miles around. One enterprising operator, noting strong gas seepages in the water of Caddo Lake, set fire to them, followed in a boat, and leased the land where the shoreline stopped the fiery trail.

Neighboring towns could use only a fraction of the gas. The waste reached such startling proportions that Caddo's burning gas was estimated to equal one-twentieth of the nation's gas consumption. Alarmed, the state made it a criminal offense to permit a well to burn or blow into the air. The Federal government refused to lease public lands in the area.

When Mike Benedum and Joe Trees arrived, Caddo had produced only 50,000 barrels of oil from eight wells during the entire preceding year. This field defeated the drillers. There would be a satisfying gush of oil for a few hours, then a violent eruption of gas, killing the oil flow. The well would usually blow out, sometimes forming a huge crater into which the drilling equipment would fall. The holes of some wells would collapse under the tremendous pressure.

The very unusualness of the factors was a tonic to Benedum and Trees. There was obviously oil here, and a great deal of it. Perhaps they were a match for the situation. Their enthusiasm may be gauged by the amount of land they leased —130,000 acres—almost three times as much as they had leased when they made their big decision in Illinois.

They drilled several shallow wells, getting some small production that could not begin to pay back the $100,000 the play had cost them so far. Then they decided to do what they should have done in Oklahoma—drill deeper. When the next hole was only 6 feet deeper than the deepest well anyone had yet drilled in this field, Caddo's first gusher began flowing over 3,000 barrels a day and another great oil boom was born.

While the tide of outgoing oil seekers was rushing back to Caddo, Benedum and Trees realized they had not matched the situation after all. Their next two wells blew out and were ruined.

In the Caddo field, as in all oil fields, oil, gas, and water had collected in layers in a trap. Gas, being the lightest, rose to the top pores where an impervious layer of curved or peaked roof rock, stopping it, formed the top and sides of the trap. Beneath the gas, the rock pores were filled principally with oil, with some gas in solution. Pressure of the heavier bottom layer of water held the oil and gas in the trap, forming the bottom of the trap. In most traps, when the roof rock is punctured, the mingled gas and oil gush to the surface just as a warm carbonated soft drink fizzes out of the bottle when the cap is removed. There was an enormous amount of free gas collected in the top of the Caddo trap, and the formations above the roof rock were soft. The tremendous pressure of the first great rush of gas would tear the hole so that the soft rocks through which it had been drilled caved in. There was no way to force the gas-cap to expand in the oil zone and provide the necessary energy to bring the oil to the top.

Benedum and Trees had never encountered a situation like this before. However, they had drilled a West Virginia well whose sides caved in. Joe Trees' father had suggested they pour cement down the hole, let it harden, then drill through it, creating their own strong wall through the soft formations. Joe Trees thought the same idea might work in Caddo. They drilled to a point just above where the bit had penetrated into the gas and oil zones in their other wells. Then they poured in cement and resumed drilling when it was hard.

They were both on the derrick floor when the well blew in. Joe Trees had a lighted cigar in his hand, but he moved faster than the gas. With the gas came oil—100 barrels an hour. In a few days they were able to control the flow. The hole held. Joe Trees had matched Caddo.

Benedum and Trees elatedly began to develop their acreage, only to discover that by solving a problem they had created others. Northern Louisiana is lake and bayou country. The partners had not bothered to lease any of these waters simply because the idea of drilling in water had not occurred to them. Gulf Oil and others hit on the new idea when they frantically looked for acreage not under lease to Benedum and Trees. The partners were startled to see derricks being erected in waters on and adjoining their leases. As these wells started to produce they found themselves in a similar situation to Mike Benedum's classic West Virginia duel with Standard Oil. But here there were dozens of opponents. Furthermore, these wells were deeper and more expensive to complete than any they had ever drilled.

When Benedum and Trees had 800,000 barrels of oil in storage and could produce as much or more every month, they again locked horns with Standard Oil. Standard offered to buy the partners' oil for 39 cents a barrel, whereas it was selling its own Caddo oil for $1.40 a barrel. Refusing the ridiculous price, Benedum went into action. He contracted to sell oil in Beaumont and to cotton mills in Louisiana for 70 cents a barrel, thereby drawing Standard's customers. He built a pipeline from the Caddo field to the railroad and triumphantly began shipping oil in tank cars. But only for a few weeks. The railroad told him they had no more tank cars available.

Mike Benedum toured the railroad's lines, counting tank cars not in use. When he learned a Standard director was also a director of the railroad he appealed to Washington. Federal court action to dissolve the entire Standard Oil combination as a conspiracy in restraint of trade was then in full swing. Benedum's plea for help received immediate, if informal, action. The railroad recounted its cars and found a mathematical error in favor of Benedum. Shipments resumed.

The partners now planned a campaign to get into a favorable bargaining position to sell their Caddo properties to Standard. A game of business began, played according to ancient rules. Benedum and Trees were to play it so often with Standard that Standards historians would later pay tribute to "the significant, if somewhat indirect and ill-defined" part they played in the company's history. "The economic service" Benedum and Trees performed for Standard "in shouldering the heavy risks of exploration was an important one. They found the oil, sold out to development companies such as Louisiana Standard and Carter, and moved on to new territories. The bonuses they received were modest when compared with the value of the oil eventually taken from the fields they discovered, but the high cash payments and early withdrawal from the proved fields appear to have been more to their liking than the routine tasks of producing, refining and marketing."

The game, of course, was in seeing how high Mike Benedum could get Standard's cash payments. Like a ruffed grouse mounting a log and drumming its wings to challenge a rival, Benedum first announced that he and Trees were going to build a refinery in Caddo. They bought land and hired experts to draw up plans. Benedum then called on prospective customers, offering oil and products below Standard's prices.

An answering drumming came from a distance. Would Benedum meet in Ohio with Commodore J. C. Donnell, president of Ohio Oil Company, the Standard subsidiary that had bought some of the partners' Illinois properties? Well, no—Benedum was too busy to go there or any place else that was suggested. Donnell arrived in Shreveport asking for a sixty-day option on the Louisiana properties. Benedum refused, but finally gave him a fourteen-day option to buy them for $10 million.

Standard requested a conference in New York. Benedum told Joe Trees and a friend, Harry Grayson, who had participated in the venture, that he wanted a written agreement from them to let him do the negotiating. "I'm afraid you boys will be too soft-hearted when you hear the hard-luck stories that will be told by those multi-millionaires," he explained.

The contestants gathered in the office of Standard vice-president A. C. Bedford, who made the first move—an offer of $2 million. "Come on boys," Benedum said, "let's go back to Pittsburgh. We are too busy to waste our time discussing such a ridiculous proposition."

Bedford suggested they talk things over. Benedum explained that they weren't anxious to sell, but would consider $10 million. Bedford countered with an offer of $3 million. Again Benedum rose to leave. This time his partners took him in the next room and urged him to accept. Benedum held them to their agreement, assuring them he would get no less than $5 million. He then spoke stirringly to Bedford of the 50 million barrels of oil he believed Caddo would produce, little realizing it would prove to contain more than six times that amount. Bedford upped Standard's offer to $5 million.

To the consternation of Joe Trees and Harry Grayson, Benedum still held out. He demanded to see John D. Archbold, Standard's president since Rockefeller's retirement.

When Archbold arrived, Mike Benedum said he would be willing to accept $5 million cash and another million after Standard had produced $5 million worth of oil. He then handed Archbold a box of products refined from Caddo oil, a notation on the cost of refining, and a price list of what Standard would receive for each product. Here was tangible proof that Standard could afford to pay the extra million. The game was over. The deal was made.

Each contestant felt victorious, and with justification. Mike Benedum and Joe Trees were free to move on to new territories. They had sold only their oil rights; the gas rights would eventually pay them many millions of dollars more. Standard, for its part, had bought the heart of one of

America's greatest oil fields for a small fraction of what it would profit them. One well paid back half the purchase price.

Much later, Benedum declared that if he and Joe Trees had developed their Illinois and Louisiana fields themselves, they would have made a billion dollars. But they were not the sort of men who would trade freedom for money. Let tho great merchants chain themselves to their desks with routine tasks of producing, refining, and marketing. Mike Benedum and Joe Trees were explorers, restlessly eager to move on.

The nation had climbed to another plateau in its development in the years between 1904 and 1909. In a population of 90 million, 200,000 Americans owned automobiles. Those who didn't now believed they, too, could buy freedom, for Henry Ford announced in 1909 he would henceforth produce his Model T—a car "built for the multitude."

The same year New York saw its first airplane. Wilbur Wright rose from Governor's Island in his awkward flying box and circled the head of the Statue of Liberty while all the craft in New York harbor piercingly cried a salute to America's realization that man was finally as free in the air as he was on the ground.

In Texas, California, Illinois, Oklahoma, and Louisiana, oil was now flowing in the quantity necessary to make the automobile and air age possible. The new discoveries were producing three times more than the old Oil Region.

The young men's visions, so dependent upon each other for success, had finally merged in glorious fulfillment. As always, fulfillment was prelude to expansion. The dreams made visible were ready to be appropriated by everyone.

There would soon be such a breath-taking demand for wheels and wings that the oil explorers would be challenged as never before to discover the prodigal provision of power to keep them moving.

9 : *A golden lane*

CHARLES CANFIELD and Edward Doheny had stepped out on faith again in Mexico. They were building a 65-mile pipeline through the jungle to the port of Tampico and had not even drilled to see if there was oil on the property they were counting on to fill the line.

"As we knew when we first saw Ebano" (the name given to their first discovery at Cerro de la Pez) "that we were going to develop oil there in commercial quantities, so we now had faith in the productivity of this new district," Doheny explained simply.

Nine years had elapsed since Canfield and Doheny had looked at their first oil spring. Now they were desperately in need of money to prove that this region would produce. Like the Amazonian tree frog that feeds on its own tail until it is strong enough to forage, the partners had been feeding their Mexican dream with the profits of all their past discoveries.

Four years after their first discovery, and after spending $3 million, they had persuaded the Mexican Central Railway to begin buying their oil. A contract to sell 6,000 barrels a day only stimulated them to buy more properties and make more elaborate plans.

They had drilled no big gushers, in the sense Spindletop

had given the word. Their biggest discovery, in 1904, flowing 1,500 barrels a day, was exciting, and confirmed their
confidence, but also increased their storage problem acutely
since they yet had no market.

In 1906, there had been a hint of what they were expecting
to find, but it happened to someone else. They were no
longer exploring Mexico alone. Sir Weetman Pearson, who
had promoted the Tehuantepec Railway across the Mexican
Isthmus, had been intrigued by his engineers' reports of oil
seepages. After the Canfield-Doheny discovery he organized
a British company to look for oil on the coast south of
Tampico. When Pearson's Dos Bocas well blew in, it caught
fire and formed a burning crater covering several acres.
Defying all efforts, including those of the Mexican army, to
extinguish it, the fire burned for forty days. By night the
great conflagration could be seen as a mysterious beacon
light by ships many miles out in the Gulf of Mexico. No
one could estimate how much oil was consumed.

The Juan Casiano property from which Canfield and
Doheny were optimistically building their pipeline in 1910
was only a few miles south of Pearson's crater. While they
were erecting pump stations, boilers, and receiving tanks,
they began drilling their first well. Oil soon began flowing
into the waiting tanks, the rate gradually increasing until at
the end of ten days the well was producing 15,000 barrels a
day. The storage tanks were filled. Canfield and Doheny shut
in the well, awaiting completion of the pipeline.

They were also anxiously awaiting the arrival of an important visitor, one who might help them solve their financial
problem. This, since the new discovery had now to be developed and marketed, was more acute than ever. The visitor
was Joe Trees. Canfield and Doheny hoped to get Trees and

Benedum to invest in their venture. The two greatest wild-catting teams in oil history were about to meet.

The terms *wildcat*, describing a well exploring for oil in any area where it has not been found, and *wildcatter*, describing the explorer, go so far back no one is quite sure how they originated. One story has it that early-days drillers, working in the hills, said they were drilling "out among the wildcats," but before the time of Drake's well, of course, the term was used in connection with unsound banking ventures, land booms, and stock schemes. Its usage in the oil business may have begun with these implications of recklessness and risk, but in oil the term never had questionable or disreputable connotations.

It is logical to assume that the quickness with which oil-men adopted the word, with its specialized meaning for them, was because the characteristics of the animal itself were those required by a hunter of oil—audacity; patience in stalking prey; sight so acute that the ancients believed a wildcat could see through a stone wall; and above all, a fierceness and independence that defies taming. "Wildcatter" was used with pride, just as the greatest compliment that could be paid a man in frontier days was to say he could lick his weight in wildcats.

Canfield and Doheny. Benedum and Trees. These men were living definitions of wildcatters at their greatest.

When Joe Trees and his Pittsburgh friends who had arranged the meeting arrived at the Juan Casiano camp, the second well was drilling. Joe Trees was as overcome as Canfield and Doheny had been ten years earlier with the strangeness and difficulties of hunting for oil in jungle so dense that sunlight could not penetrate it. But now there was more than a bubbling oil spring to make a believer of him. His first morning in camp, Trees went out to check on the drilling.

The well had not yet reached the depth where oil had been found in the first well. It seemed that there was nothing to be expected except the eternal, hypnotic fascination of watching and listening to the rhythmic operation of the tools pounding unseen rocks far below.

Doheny explained that as soon as they reached within a few feet of the depth of the first well, they would set pipe in the hole so that the flow of oil could be shut in from the surface. There was no tankage—their tanks had been filled by the first well—but when the pipeline was completed they would have plenty of oil ready to flow into it.

Trees did not have to use his imagination in determining what "plenty" would mean. There began a familiar ominous rumble, but all Trees's past experience had not prepared him for the sight of the great geyser that followed. It soon turned the jungle a glistening black for miles around. Telling his partner about it later, Trees could only declare earnestly, "Mike, you've never seen an oil well. Those things we drilled in Illinois and Caddo were just little creeks of oil. I tell you these Mexican wells are oceans."

The Canfield-Doheny ocean blew in uncontrolled at the rate of 70,000 barrels a day, and for nine years this one well alone produced all the oil that Canfield and Doheny could market. When it ceased flowing, it had produced 80 million barrels—or eight times as much as the average United States *field* will produce today.

Mike Benedum, faced with his partner's enthusiasm, felt that perhaps he and Trees should go to Mexico on their own instead of joining forces with Canfield and Doheny. However, they compromised by buying half a million dollars' worth of the Mexican company's stock with an option to buy an even bigger block later.

Mike Benedum and Edward Doheny were too much alike in certain respects for their combination to last long. They were each the aggressive leader of their respective partnerships, and they were fiercely independent types. Both were shrewd bargainers, though Benedum had the edge in experience.

As soon as the pipeline to Tampico was finished Doheny suggested that Benedum approach Standard Oil. Next to finding oil, Mike Benedum's greatest pleasure was just such a trading contest. In New York, he talked Standard into agreeing to buy 12,000 barrels daily at 52 cents a barrel, and he threw in a free shipload of oil to test.

To Benedum's surprise, Doheny was vehemently critical of the deal, saying he had already talked to the Mellons and was sure he could do better with Gulf. Benedum was so annoyed he decided to sell his Mexican stock. Trees agreed. Benedum advised Standard the Mexican sale was not satisfactory.

Doheny, unable to conclude negotiations with Gulf, wanted to try Standard again. Benedum told him to handle matters direct. "Maybe you can do better than the fifty-two cents I got," he said. He was amused when Doheny's trading resulted in accepting Standard's offer of 39 cents a barrel.

Benedum and Trees sold their stock in the Canfield-Doheny company. With the half a million dollars' profit, they went into Mexico on their own. They couldn't be satisfied with someone else's gushers. Soon they had one of their own, opening another rich new oil field. Mexico was big enough for both partnerships and dozens of other companies, too.

As one gusher after another roared in, the Mexican coast became known as the Golden Lane. The importance of these great gushers was not so much that they rocketed Mexico into the position of the world's second-biggest oil-producing nation; rather, it was in the profound impression they made on American oil explorers. If there was an abundance of oil in the earth of Mexico—so little foreseen—why not also in some undrilled spot in Texas or Oklahoma or whatever other state a dreamer of oil might happen to enter?

10 : *The mystery well*

Harry sinclair looked stonily at his lease man. "So, you came back with a fish eye in your pocket, did you?"

"Mr. Sinclair, you don't understand what happened. I—"

"Damn it, I'll be the judge of what I can understand, young man. Did you get any leases?"

"No, sir." The lease man was flustered. Sinclair could always make you feel that somehow you were wrong, that you hadn't done your best. "You see, sir, Slick and Shaffer roped off their well on the Wheeler farm and posted guards and nobody can get near it."

Sinclair said nothing.

Gathering courage, the lease man rushed into his story. "I got a call yesterday at the hotel in Cushing from a friend who said they had struck oil out there. A friend of his was listening in on the party line and heard the driller call Tom Slick at the farm where he's been boarding and said they'd hit.

"Well, I rushed down to the livery stable to get a rig to

go out and do some leasing and damned if Slick hadn't already been there and hired every rig. Not only there, but every other stable in town. They all had the barns locked and the horses out to pasture. There's twenty-five rigs for hire in Cushing and he had them all for ten days at four-fifty a day apiece, so you know he really thinks he's got something.

"I went looking for a farm wagon to hire and had to walk three miles. Some other scouts had already gotten the wagons on the first farms I hit. Soon as I got one I beat it back to town to pick up a notary public to carry along with me to get leases—and damned if Slick hadn't hired every notary in town, too."

Sinclair laughed. This was the way the game was played. "It looks like they think they're going to make a killing."

"Yes, sir. Slick has leased everything solid in a radius of six miles from the well, except some Indian leases. I've been checking the records and you have to get the Interior Department to put them up for sale."

The lease man spread out a map. As Sinclair studied it, he snapped, "Did you find out anything about the well?"

"Yes, sir. One of the drillers came into town last night and I bought him some drinks. He figured it was all right to talk to me because Slick and Shaffer have got all the leases. He says it's about a forty- or fifty-barrel well as she stands now. They're down twenty-two hundred feet and only about two inches into the oil sand.

"Now that's not much of a well when the nearest production is sixteen miles from here. Grabbing up all that land before they know how thick the sand is or how rich it is, looks pretty risky to me."

"That'll be all," Sinclair said. "I'll let you know what to do."

"Tom Slick sure tried to plug all the loopholes," the lease man went on. "His driller told me that when they called him, he was so mad, knowing everybody on the line would be listening, he tore the phone off the wall. But he wasn't half as mad as the old lady who didn't know what was going on and couldn't find out with her phone gone."

"You could take a few lessons from Mr. Slick," Sinclair said. "Next time, have your own rig and notary under contract."

After the chastened lease man left, Harry Sinclair brooded over the map. A 40- or 50-barrel well in this Oklahoma wasteland wasn't much to get excited about. It was a wild wildcat. Maybe the grandstand play of making it a mystery well, and hiring all the rigs and notaries and snapping up the leases, was just an attempt to create an unwarranted market.

Yet this young fellow, Tom Slick, had promoted other wells in the area. He had a small-town banker, B. B. Jones, as a partner, and he and Jones had gotten C. B. Shaffer, a wealthy plunger from Chicago, to back this new play.

Harry Sinclair remembered Tom Slick. He had come into Sinclair's office to sell some leases: a gaunt, cadaverous young man with a shock of snow-white hair and dark, deep-set eyes that had almost a mystical fire in them.

Tom Slick's father had been an old-time oilman in Pennsylvania and young Tom, starting as a lease broker there, had gone to Illinois when Mike Benedum opened up the Robinson field, and had followed the oil tidal wave to Kansas and Oklahoma. An exceptionally quiet, modest fellow, he had stated calmly, when Sinclair needled him, that he had gone into the oil business because he was going to make a million dollars. Of course everybody else who went into it said that, too, but the way Tom Slick said it was different. Harry Sinclair could recognize the difference.

It made him think of a rainy afternoon in his office in Independence a few years earlier. Everyone had left except Billy Connelly, who didn't know the meaning of time when it came to working for Sinclair.

"Billy, how old is Rockefeller?"

Connelly looked startled. "Oh, I don't know. He's an old man. He must be about seventy."

"Well, when he dies, I'm going to be the biggest oilman in the world, bigger than Rockefeller."

"Harry you're crazy. Nobody can be bigger than Rockefeller."

"I can. He hasn't got anything I don't have. He hasn't got any more brains than I have. And I have something he doesn't have—youth, and time ahead of me. You'll see."

Harry Sinclair knew the difference between vague ambition and that dedicated, driving force that was like the steel barrel of a gun, preventing the shot from scattering.

Perhaps Tom Slick had a hunch about these steep, scrub-covered hills and deep ravines 40 miles west of Tulsa. The land was worthless—unless Slick's million dollars lay concealed beneath its surface. Hovering over his mystery well, Slick seemed to feel the way Sinclair had once felt about a 160-acre lease in Oklahoma. Two dry holes had been drilled on it, but he had remained so sure there was oil present that he instructed Connelly to buy the lease for $50,000. When Connelly protested that the owner would be glad to get rid of it for $5,000, Sinclair sternly insisted on offering $50,000— there was to be no question about getting it. Sinclair had thirty producing wells on that lease now. The two dry holes were the only two places where there wasn't any oil.

To find oil you had to sense that it was there. You played trends, but you also played hunches. Harry Sinclair liked cemeteries. He also liked places where plenty of blackjacks grew. Tom Slick had drilled his well in real blackjack country.

Sinclair had already made his million in Glenn Pool and in leases in smaller fields. What he needed now was something big to catapult him nearer his goal.

In addition to his youth, Sinclair had something else that Rockefeller didn't have—the conviction that the way to greatness was in the ownership of great oil reserves. Rockefeller had begun by scoffing at ownership of the sources, sure that his monopoly could be sustained by control of refining, pipelining, and marketing. So much oil had now been found that Rockefeller's Trust could not control it. His type of enterprise could not yoke so many oxen. As a matter of fact, the accelerating oil discoveries had begun to break Standard's monopoly even before the Supreme Court, in 1911, ordered the trust dissolved.

Busting the Trust was an ironical joke that had backfired on rampaging President Theodore Roosevelt. He had forced the Trust to break up into thirty-eight companies, but with what results? They still had the same owners, and as yet they were not competing with one another. For the first time—the result of the action—the public was aware of how valuable Standard's holdings were. All the companies' stocks jumped in price. Rockefeller was a hundred per cent richer after the dissolution.

Standard Oil itself had finally realized, before the government began its prosecution, that the real threat to its monopoly came from not owning its raw material. Still averse to risking money hunting oil, it had begun buying most of the good discoveries.

Sinclair saw clearly that if he was to become greater than Rockefeller, he must found his empire on ownership. Until he, too, had the money to buy other gamblers' discoveries, he must continue to be one of the greatest gamblers. What he had to do now was find oil in such quantities that he would not have to sell to Standard. He remembered an interview Rockefeller had given during the height of the Trust's battle with the government. At the urging of associates he had reversed his ancient policy of telling the public nothing, allowing a New York *Times* reporter to trot along beside him as he played his daily golf game on the private links of his Pocantico Hills estate. Dressed in a short silk kimono, lined with paper, and wearing a high helmet, like an inverted scoop, lifted from his head by spokes attached to a hoop, Rockefeller looked "for all the world like a Chinese mandarin," the reporter noted. "Only the coral baton was missing."

Rockefeller genially defended Standard's philanthropic motives, but what had sparked Sinclair's interest was the magnate's almost defiant expression of the core of his belief. "We have come to a new economic era," he said. "In the future, business is going to be carried out more and more by aggregations of capital. It cannot be otherwise. The day of individual competition is past and gone."

Harry Sinclair agreed—part way. The old man was right about the role of capital; he was wrong about individual competition. Standard, in Sinclair's view, offered a pattern of an efficiently controlled aggregation of capital that individuals could use in forming other aggregations to provide competition—individuals like Harry Sinclair.

Again he looked at the Creek County map the lease man had given him. The Indian leases, outlined in red, were scattered through the big circle of leases Tom Slick had so cleverly monopolized for himself and his associates. Sinclair had that familiar physical awareness of rising excitement, his heart beating a little faster, the pit of his stomach suddenly becoming alive and hollow. If Tom Slick thought so highly of the area, perhaps he was even now asking the agent of the Department of the Interior to put up the Indian leases for sale. If so, this was a game two wildcatters could play.

For several weeks the Slick and Shaffer mystery well was the principal topic of conversation among Oklahoma and Kansas oilmen. The enthusiasm that any new discovery generated was missing, for there seemed no chance to lease any land around it. However, in hotel lobbies, offices, and bootleg joints there was endless speculation as to what the well had found.

When the discovery well was drilled 2 inches deeper in the sand, its production increased to 150 barrels a day, but a second well produced only 25 barrels. The number of those who argued it wasn't much of a find increased. There were too many wildcatters who could recount the bitter disappointment a man feels when his bit reveals the thinness of the oil sand he has finally hit. It took oil sands 100 feet thick, like Glenn Pool, to make fortunes.

A month after the discovery, it was announced that Harry Sinclair was highest bidder for the Indian leases. Oilmen's reactions to the prices he paid were a mixture of amazement and ridicule—$8,000 for one 40-acre lease, $7,350 for another. All told, he paid an average of $200 an acre. Slick had paid a dollar an acre for his. The *Oil and Gas Journal* summed it up: "The prices paid for the property were greatly in excess of what the land is really worth."

Everyone was surer than ever when Sinclair's first well on

his expensive leases was a dry hole. The Cushing field was soon almost forgotten. The other wildcatters pushed off to other parts of the state. Sinclair, however, kept on drilling, and each new well gradually extended the producing area. Eventually he had the satisfaction of bringing in the field's biggest well—a 1,500 barrel gusher.

By January, 1913, 25,000 barrels a day were flowing from Cushing wells. There was no question now but what Cushing was a big field, although it certainly was no Glenn Pool. Then—the wells declined. It had been flash production from a shallow depth. In the hotel lobbies Cushing was called "a dead one."

It was not until a year and a half after the discovery well that the oil industry suddenly realized that Cushing was, after all, a giant oil field. No other great oil boom ever took so much time being born.

While the drillers were blindly probing all around, the Cushing discoveries had moved northward, until oilmen were surprised to see that the field was already 8 miles long and 2 miles wide and the end was not in sight. What was more, it had extended into an area where leases were available.

In November, 1913, a test well was drilled 300 feet deeper than any of the producing wells in the field. It found a new pay horizon—the prolific, thick Bartlesville sand. Wells gushed 4,000 and 5,000 barrels a day.

The great rush began.

So many wells were started, there was a shortage of drillers. As the news traveled over the grapevine, some four thousand workers began to arrive from Texas, Kansas, Illinois, Indiana, Pennsylvania. And for every worker, seemingly a dozen boom-town parasites followed on his heels.

Tent cities mushroomed the length of the field. Drum-right, 10 miles from the town of Cushing, was the hub of activity. A one-street town, hastily slapped together in a gorge, its flimsy two-story buildings rose like stairsteps up the hillsides from Tiger Creek. Gamblers, highjackers, saloon owners, and brothel keepers poured into town. Beds were never cold, their occupants grudgingly relinquishing them to the next weary worker every few hours around the clock. Men slept on boards in tarred-roof shacks, in tents, Indian teepees. They fought over places in line in order to eat tough meat, soggy potatoes, and pie with celluloid crust. One entrepreneur, arriving with a $42 stake, opened a restaurant not much bigger than a cigar box. Serving clean, well-prepared food, he made a $500 profit in a week. He eventually made more of a fortune from the oil in his cook stove than the majority of those who were bringing it up from the ground.

Drumright had an official population of 1,500, but 6,000 got their mail there. The post office was the only pool hall in town. The daily mail was dumped on the pool table and pawed over by each shift coming from the fields, until late-comers could scarcely read the addresses on the oil-stained envelopes.

Not an inch of the 10-mile road from Cushing to Drum-right was ever empty, day or night. A never-ending stream of 250 teams of wagons and horses hauled boilers, smoke-stacks, cables, tanks, steel pipe, and lumber for derricks and stores. Horses' noses touched tailgates of wagons in front of them. It was the noisiest 10 miles in America—a pandemonium of cursing, shouting drivers, cracking whips, groaning wagons, clanking machinery. On hot days horses would drop dead under the strain and there would be a seething riot of struggling men and animals. There was no room for automobiles. A second road was hastily constructed. When it rained, this road, too, was crowded—with mired, abandoned cars.

F. S. Barde, the Oklahoma correspondent of the Kansas City *Star*, caught up in the vortex of this astonishing boom, reported enthusiastically that "any man with a gill of red blood in him is stricken immediately with the oil fever upon reaching Drumright. He begins seeing things grow larger and larger at every turn. He is soon talking in terms of thousands where a day earlier the best he could do was about two bits. The atmosphere is charged with superlatives and hyperbole. It is bad form to accuse a man of lying, because a moment later he may make good. There is a smell of crude oil and gas at Drumright that gets into a man's nostrils and causes a delightful inebriation, the symptoms tending toward a magnification of wealth lying around loose in the country."

There was no magnification. If anything, everyone was looking down the small end of the telescope. Cushing would one day place twenty-first in the list of America's biggest oil fields. It covered 32 square miles and contained 455 million barrels. Only one other field in Oklahoma would surpass it.

Wooden derricks quickly supplanted Sinclair's blackjacks in hills and ravines. From an ebb of 25,000 barrels a day the ocean of oil rushed back to a high tide of 300,000 barrels a day. Never since the palmy days of Glenn Pool had there been such opportunities for money making; and never had there been such violence, crime, and tragedy.

The characteristic phenomena of the oil-boom towns were chiefly due to difficult transportation and bad roads. When these transient communities produced an abundance of fuel energy and wealth, an abundance of swift transportation and good roads resulted. Oil field workers could then easily commute to a new discovery area from their homes or establish their families in a nearby settled community where the roots of social order had already gone deep. However, in the turbulent early days, the workers were in effect staked to their wells, the length of the rope extending no further than their legs could carry them to the wretched hellholes of the nearest "town."

As the mammoth Cushing field spread, a whole series of settlements erupted like boils down the length of the field. In mocking tribute to Drumright, which took its name from Aaron Drumright, owner of the farm that had yielded such an unusual crop, the names of the later towns all ended in "right"—Dropright, Gasright, Alright, Downright, and Justright. When Tiger Creek was dammed to catch a fresh overflow of oil, there was a new one—Damright.

Only those who were capable of meeting the grueling physical demands of working on a cable tool rig in an early boom field could understand the psychology of workers and their capacity for whisky, which the next day's work would sweat out of them. A typical roustabout once scribbled his schedule on the back of an envelope:

11:00 A.M.	Get up
11:00—11:30	Sober up
11:30—noon	Eat
Noon to midnight	Work like hell
Midnight—3:00 A.M.	Get drunk
3:00—3:30	Beat hell out of them that's got it coming
3:30	Go to bed

With no place else to go, nothing else to do, the workers spent their pay as soon as they made it. There was always the next day's wages and the vague remembrance of their original idea, which after the first shot of whisky assumed certainty, that after they had worked long enough they would save up enough money to buy a rig or a lease of their own and become an oil millionaire.

These men were more reckless than lawless. As F. S. Barde observed, "in Drumright, by turning your head, you can see a dozen fights in one moment. The fighters were allowed to fight to a finish as it was a waste of time to mix in other men's quarrels. You might have one of your own if you became 'auspinarious,' an Oklahoma word meaning 'buttinskism.' "

Recklessness reached dangerous heights when wildly hilarious groups would stagger down the main street shooting for fun into the pine shacks. Those inside would quickly fall flat on the floor with no thought of complaining afterward. More faces were scraped in this fashion by rough pine boards than by barbers.

Tempers flared cruelly in this atmosphere at any real or imagined offense. On one occasion, the obvious respectability of a young tool dresser in Dropright's pool room attracted the drunken attention of Blackie, a notorious gambler, who asked him to have a drink. The young man refused.

"Too proud to drink with common folk, huh?"

"I'm not proud, but I'm not drinking just now."

Blackie forthwith shot him in the thigh.

Workers and professional outlaws had no patent on violence. In a hotel lobby, an attorney saw a lease man with whom he had been disputing the ownership of a lease. In sudden rage at the sight of him—the man was quietly registering at the desk—the attorney picked up an iron stand and smashed his skull.

Some murders were committed with only the murderer's

knowledge of them. Many years later when production declined and the huge 55,000-barrel steel storage tanks were
dismantled, oily skeletons at their bottoms finally revealed
the secret. Perhaps these were the answers to the pathetic
classified advertisements that regularly appeared in the *Oil
and Gas Journal* during the height of the boom.

From El Paso, Texas: "Mother wants address of C. E.
Sloan. He was last heard of in the Cushing field in September, 1914."

From Mansfield, Louisiana: "Any information pertaining
to Moncel Middaugh, who has been missing from home since
June, 1914, will be greatly appreciated by a heartbroken
mother and father. Moncel is 18 years old but could easily
pass for 21 years. Of average height, about 5′ 9″, weight
about 160 pounds, dark brown hair and gray eyes."

Or perhaps the answer to this advertisement did not lie
in the bottom of an oil tank, but in a typical newspaper
notice that a young driller had committed suicide, leaving
no knowledge of who he was, only a note that he was "sick
and tired of a hopeless life."

Or perhaps the despondent suicide was the young man
sought by an attorney in Olean, New York: "Wanted. Address of William T. Beattle, lease hand, to settle estate in
which he inherits considerable amount."

Violent death stalked a man at his work as well as during
his leisure hours. Nitroglycerin, used in the wells to shoot
open the oil sands to increase their flow, accounted for as
many deaths at the wells and on the roads the nitroglycerin
wagons traveled as did the explosive mixture of whisky and
gunpowder.

Little care was exercised in the control of gas escaping
from the wells. On damp nights it would collect in pockets

in low spots on the roads. A lantern, a cigar, a spluttering automobile headlight—and Cushing field would receive another sacrificial offering. These fatalities became so common that signs were posted on the roads— DANGER: GAS POCKET AHEAD.

A drilling crew did not have to leave the rig to lose its money or find its lives endangered. Accommodating high-jackers were regular and expected visitors to the wells. A six-shooter became as necessary a rig tool as the sledge hammer used to dress the bits.

Matt Kimes, one of Oklahoma's most famous bank robbers, learned his trade at Cushing. "Matt started to work for us on a lease," relates Tommy O'Rourke, "but he was a third generation bank robber just like I'm a third generation cable tool driller and he couldn't do well at anything else."

O'Rourke's grandfather Ed, and Ed's two sons, Ed and Tom, had been expert cable tool drillers in Pennsylvania, and like others of their rugged breed were the sturdy, realistic bedrock of the Oklahoma and Texas booms, as capable performers in their profession as Harry Sinclair and Mike Benedum were in theirs.

Men like the O'Rourkes were more than a match for Cushing's lawlessness. Ed O'Rourke, Jr., went into a saloon one night and soon found himself encircled by a menacing group of outlaws.

"We understand you tipped the law that we've been doing some bank robbing," the leader stated grimly.

"I sure as hell didn't, and I'll prove it to you," O'Rourke flashed. "I'll take you bastards on two at a time and lick every one of you."

Nobody needed a further invitation. It was a classic fight. O'Rourke won, although he took such a beating he could scarcely be distinguished from his defeated opponents. A week later he was approached by one of the gang, who, as O'Rourke squared off, hastened to assure him he had come to discuss business.

"We found out you wasn't the one," he apologized. "We're sorry we caused you so much trouble and we want to pay you for it."

O'Rourke refused to accept any compensation.

"Well, we'd sure like to have you join our gang," the bank robber said admiringly. "You'd make more money with us than drilling wells. A man with your guts would make a good bank robber."

O'Rourke acknowledged the compliment but declined the honor.

Conditions in Drumright and all the other "Rights" finally became so intolerable that the oil producers stepped in to do the work of the inadequate law-enforcement agencies. The whisky was hurting their business. Prohibition had been one of the conditions imposed by the Federal government for Oklahoma's statehood. It was supposedly a measure to protect the Indians, but it was never any more successful than national prohibition was later on. The producers hired four U.S. marshals and put up a jackpot of $5,000 to be paid if they dried up the liquor sources. For a few days the boom towns were whiskyless, and a justice of the peace and a constable dared to move in.

Liquor selling went underground. The rowdyism was only curbed, not stopped, but the oil producers were as pleased as if they had founded the WCTU. It made the boom manageable. A driller could still satisfy his thirst, but he had to do it less publicly, and often only on "Choc"—a foul beerish brew named after its long-forgotten Choctaw Indian inventor. The consumption of lemon extract reached an astonishing figure, the sudden demand for it in Oklahoma baffling its manufacturers in more civilized states.

The oil producers recorded the exuberant emotions and abandon of the boom even in the names of their companies: the Midnight Oil Company, the Invincible Oil Company, the Future Oil Company, the Ideal Oil Company, the Only Oil Company. When one producer discovered there was already a Clover Leaf Oil Company, he settled for the Lucky Leaf Oil Company. The Success Drilling Company, the Black Panther Oil Company, and the Hoppy Toad Oil Company were there. The River Shannon Oil Company, with Pat Reiley as manager, had headquarters at Shamrock, a few miles from Cushing.

At the height of the boom, that wily old wildcatter John Galey, who had been operating alone since his partner, Colonel Guffey, bought his interest in their Spindletop discovery, came to see the excitement. Cushing was Galey's dream come true for other men. In 1893 he and Guffey had leased all the land between Glenn Pool and the Kansas line, in addition to 1,800,000 acres in Kansas. All the Cushing field had once been theirs. If only they had drilled a well!

"When I consider what might have happened if we had held on to all of our holdings it staggers me to think how rich we would have been," Galey said as he toured the field. "But I have no regrets. It is the fortune of the game and I am glad to see so many of the producers making good and am glad to know that the percentage of dry holes is so small."

He ruefully revealed that since Spindletop he had been hunting for oil in West Virginia, Indiana, New Mexico, Wyo-

ming, Utah, and Mexico "without increasing my balance at
the bank to any appreciable extent."

Galey had been nineteen years old when the sight of the
Drake well had awakened the explorer in him. Like Mike
Benedum, this intense little Irishman was happy only when
he was exploring new territory, but he lacked Benedum's
financial astuteness, and the profits of his vision generally
went into other men's pockets. He was seventy-four when
he saw what might have been at Cushing. Four years later,
his death at Joplin, Missouri, passed almost unnoticed by
the industry. He died still exploring, still hoping, but his last
dream was not of oil. He was seeking that elusive fortune
in lead and zinc mines.

No one thought less of old John Galey when he declared,
during his visit to Cushing, that the Oklahoma field could
not be compared to Spindletop. "I do not believe that any
living man will see Spindletop's equal," he maintained.
"Why, on Spindletop Hill, I have seen as much oil go to
waste every week as is produced in the entire mid-continent
in a day!"

There was truth in what he said. Spindletop was unique.
But so was Cushing. Each of the great booms had its own
circumstances, its own personalities, its own distinctive impact on the development of the nation that made it utterly
different from any other oil boom.

Another visitor to Cushing during its peak time was effu-
sive about what he saw. Dr. Wilhelm Wunsdorff, personal
geologist of Kaiser Wilhelm, had inspected all the Mexican
and United States oil fields before his last stop in Oklahoma.

"I have never been so impressed with the possibilities of
any country as I have been in Oklahoma," the German au-

thority marveled to his gratified hosts. "It is the greatest oil country in the world. I have been to Russia, to Roumania, to the Straits Settlements and Japan, to every place where oil is produced in quantities, and I must say that this is the most productive, high grade oil country on the globe."

Ironically, Dr. Wunsdorff was praising what would be the key to his country's defeat a few years hence.

"The Allies floated to victory on a sea of oil," proclaimed Lord Curzon, British Secretary of State for Foreign Affairs, pointing out that 90 per cent of Allied petroleum needs came from America. A major part of this "sea" flowed from Cushing.

When World War I began, one of the first bombs ever dropped by an airplane on a ship was a German bomb that struck a Standard Oil tanker. The ship was the *Cushing*.

11 : "What does Rockefeller have that I haven't got?"

H ARRY SINCLAIR was preparing to take his giant step forward.

In 1915 he moved to Tulsa, Oklahoma, the little cow town that the wealth from Cushing, 40 miles away, had established as the "Oil Capital of the World." The new millionaires competed as keenly in building mansions as they did in acquiring leases in the field. Each tried to outdo the other in the lavishness of salons, the gold plate of bathroom fixtures, and the spaciousness of ballrooms.

Sinclair held himself aloof from this posturing new society. He built a handsome brick mansion on the crest of a hill overlooking the Arkansas River, but while the others were dazzled by one success, he was planning his campaign. Those Indian leases circled in red on the map were worth the millions he needed to open the next door.

Though no one suspected what Sinclair's goal was, the calculation with which he maneuvered did not go entirely unremarked. When *Sporting News* described him as the oil and baseball magnate "who is a Coal Oil Johnnie," the *Oil and Gas Journal* hastened to contend that the comparison was not apt. "Coal Oil Johnnie never expected any returns for the money he so lavishly spent," it commented. "Sinclair gives every dollar thrown away by him a return ticket."

Sinclair bought the Kansas City franchise in the newly organized Federal Baseball League and moved it to Newark, New Jersey. Even in this hobby the pattern of his ambition was apparent. This was the only serious attempt ever made to compete with the American and National leagues. It was an exciting two-year battle. Although Sinclair did not win, the *Oil and Gas Journal* had correctly predicted what would happen. When he sold all his ball players in 1916, reports that he had lost a fortune led him to announce, "The best part of it is that I have made money!"

While Sinclair had clearly viewed Cushing as a means to a goal yet to be achieved, Tom Slick's goal had been more than realized when he sold his Cushing properties for $2 million.

"Two million dollars is more than I can spend in a whole lifetime," Tom Slick said, and took a trip around the world.

But something was wrong. There wasn't the joy in spending money that Slick had thought. Could a man of twenty-six be without a goal? When he returned to Cushing and saw even more intensive activity than when he had left, the fire rekindled in his eyes. He discovered then that one goal must always be replaced by another, that achievement is not what counts, but the achieving. Accordingly, he went back in the oil business. Seventeen years later, when he sold some of his oil properties for $35 million, he was asked how soon he thought he would quit.

"When I die, or my health breaks down, or when I lose too much money at it," he replied. "But I'm not working simply for money. I've got enough. I keep at it because of the love of the game. I'm a born trader. I get a kick out of it, even when I don't win. It thrills me to know that I'm doing a man's work, producing something, and as long as I can win at it I'll keep going."

In 1930 Tom Slick quit the oil game for the first of his "whens." He died at forty-six, literally from overwork, having gone from one extreme to the other, never learning that just as no work is not the answer, so neither is all work.

To the owners of the land from which Cushing's wealth gushed forth, the event could be classified as an act of God or of the devil, depending upon the way they reacted to the sudden, unexpected riches the explorers brought them.

Frank Wheeler, a stonemason, was wearily preparing for bed one night in 1911 when Tom Slick came to his door seeking a lease. Grudgingly, Wheeler conformed to pioneer etiquette and invited the stranger to have some supper and spend the night. There was hardly enough in the house to feed his wife and nine children. Only that morning he had been refused credit for a sack of flour at the crossroads store.

Wheeler's 160 acres were a farm in name only. Mrs. Wheeler could barely manage to make the stony soil yield a few vegetables while Wheeler ranged the countryside building storm caves and cellars for the few who could afford to hire him. Wheeler felt he had come to Oklahoma too late. The firstcomers had siphoned off the cream, seized all the opportunities.

As Slick explained that he wanted a lease to drill for oil on his land and that he, as owner, would receive a royalty of one-eighth of all the oil found, Wheeler could not really follow the thought. It had no meaning for him until Slick mentioned that he would pay a dollar an acre if he would sign the lease. For $160 in cash, Frank Wheeler would have sold his farm at that point. He could scarcely wait for daylight so that he could go to town to sign before a notary for this stranger who was inexplicably ready to give him something for nothing.

When the Wheeler No. 1, Cushing's mystery well, started drilling, it was more of a mystery to Wheeler and his family than it was to the oilmen who were trying to learn what Slick and Shaffer were finding. Wheeler was still trudging the countryside in search of cellars to build. The great cash bonanza had long since disappeared.

The day Wheeler received his first royalty check for $500, not even Harry Sinclair or Tom Slick could have understood what the discovery of the Cushing oil field meant to this one man. How can the soaring eagle know what the emotions of the rabbit are when he has struggled loose from a trap?

Sudden riches are a dangerous thing. One farmer went insane when his royalty checks began to arrive. Frank Wheeler was more stable. In a few months, there were sixteen wells on his 160 acres, harvesting him a crop of $300 a day. It was just a beginning. Wheeler did what he had long wanted to do—he left his once-hated farm and moved to

Stillwater, so that his eight daughters and one son could attend the state agricultural college. As each graduated and married, the wedding present was a farm—the finest agricultural land oil money could buy. Then Frank Wheeler and his wife set off to explore the world. Who said he had come to Oklahoma too late?

For thirty-five years Wheeler No. 1 produced, until finally only a barrel a day could be pumped from it. The lease still belonged to the Deep Rock Oil Company, which had been organized by C. B. Shaffer. When Deep Rock decided to plug and abandon the historic discovery, it sentimentally erected a granite tombstone with a bronze plaque, naming the deceased "Old Faithful!" and commemorating the fact that "Its completion by C. B. Shaffer et al on April 1, 1912 set off one of the greatest oil booms in history."

In 1954, the General American Oil Company acquired the grave along with Deep Rock's livelier assets. General American's engineers decided to experiment with modern techniques. In 1956, they cleaned out and deepened Old Faithful. Then they forced 20,000 gallons of oil and 20,000 pounds of sand down the hole, a treatment that fractured the old "dead" oil sand. Old Faithful quivered, shook, and roared back to life, gushing 648 barrels of oil a day, six times more than she had originally yielded for Slick and Shaffer.

There were many owners of the 32 square miles of land under which Cushing lay. At the beginning of statehood, five years before Wheeler No. 1 was drilled, the area was officially included in Creek County, covering part of the former lands of the Creek Indian Nation.

On the map, Creek County was neatly divided in townships of 6 square miles, each township containing thirty-six sections, each section containing four quarter-sections of

160 acres apiece. Each quarter-section represented the final carving up of North America by the white man, as this was the amount each individual Indian was forced to accept as his allotment. Two fifths of the Cushing field was held under leases supervised by the Department of the Interior for their Creek owners. Incredibly, seventy or eighty of the Creeks refused to accept the oil royalty checks due them; accepting would have meant recognizing the right of the government to allot land to them—an action that violated the treaty made in 1832 when the Creek Nation was removed from Alabama to Indian Territory.

On the other hand, there were Creeks who did not cling so tenaciously to the old cause. Eastman Richards delightedly made the choice between money and principle. His royalty checks were such an asset with the ladies that a good portion of his wealth was spent in defending alienation of affection suits. He built a whole town, Richardsville, just for himself, so that he would not have to pay at his own drugstore, dry-goods store, butchershop, or bank.

Jackson Barnett was another who also accepted his arbitrary allotment in the Cushing field. He leased his land to an oil company, but a rival company succeeded in establishing him as an incompetent in the county court, had a guardian appointed, and took a lease from the guardian. His plight was not unusual. As Angie Debo comments in her book *And Still the Waters Run:* "It had always been recogognized that mentally defective Indians like other defective adults could be placed under guardianship, but it was not until about 1913 that it began to be apparent that all Indians and freedmen who owned oil property were mentally defective."

As a full blood, Jackson Barnett's oil leases and competency were subject to the Department of the Interior's judgment. When the fight between the rival oil companies was brought to Indian Agent Dana H. Kelsey for decision, he ruled that Barnett was incompetent only to the degree that he was illiterate and a full blood. The incompetency action was set aside. However, the two oil companies made a private compromise agreement to develop the lease, and the guardian was permitted to remain to guide his affairs.

When the biggest gusher in the Cushing field—14,000 barrels a day—came in on Barnett's property and he began receiving $3,000 a day in royalties, the old full blood was reported by the Muskogee *Times Democrat* to be "running wild in the woods of Okmulgee County. He roams through the woods, living on herbs and bark and what game he can collect. Relatives let him wander and always keep a supply of fuel and food in his cabin."

"Running wild" was a typical Oklahoma white man's interpretation of the simplicity of a conservative full blood's way of life.

The legal entanglements and simple-minded eccentricities of Barnett, "the richest Indian in the world," were for years a staple of the nation's Sunday supplements. He eloped with a white bride to Los Angeles, where his favorite diversion was directing traffic on the corner near his home. At one time he amiably put his thumb mark on an agreement to donate half his fortune to the American Baptist Home Mission Society. Lawsuits between the Society, guardians, the State of Oklahoma, relatives, alleged relatives, the Department of the Interior, and the Department of Justice kept him in court the rest of his life and his fortune in litigation for a decade after his death.

At the close of the Civil War, the government had insisted that the Five Civilized Tribes grant full citizenship to their Negro slaves. The Creeks had the largest number of these freedmen of any tribe; when the tribal roll for allotment was prepared the Negroes equaled the number of full bloods. Despite the Indians' objections, the government allotted Indian land to the freedmen on the same basis as to the Indians. As freedmen were not under similar restrictions, most of their land quickly passed to white ownership. However, no one was interested in certain allotments around Cushing presumed to be worthless. As a result, in 1914, ten-year-old Danny Tucker was one of the richest Negro boys in the world. The oil beneath his allotment paid him $6,000 a month.

Nearby, the allotments of two Negro orphans, eleven-year-old Sarah Rector and her eight-year-old brother, Manny, had already yielded them $300,000, and their fortune was increasing at the rate of $10,000 a month. It would become even greater when the wells were deepened to the rich Bartlesville sand. Their mother had died of tuberculosis and their father had died in the state penitentiary. When their story was published in a German newspaper, Sarah received twenty marriage proposals.

Sarah and Manny were more fortunate than two other Negro orphans whose allotments had been in the Glenn Pool. These luckless waifs were murdered in a shack mined with dynamite by a group of whites who were ready to claim their wealth with forged documents and false heirs.

With the help of conniving politicians a scanty cloak of legality was thrown over many kindred instances of ruthless plundering. A particularly clever way of stealing a full blood's oil properties was to have his restrictions removed by special act of Congress without his knowledge. This was easily done by attaching riders to Indian appropriation bills.

Meanwhile, the unsuspecting Indian would be persuaded to give power of attorney to a white "friend," who, as soon as the act was passed, would deed the property to a confederate. Or the Indian might be induced to sign a deed for a small payment, not knowing what he was signing. The mixed bloods, not being restricted, could be dealt with more summarily. No questions at all were likely to arise over the disposition of their allotments. One mixed blood who owned a rich allotment in Glenn Pool was plied with whisky until he was in suitable condition to sign away "voluntarily" the deed to his allotment. A week later he was found beaten to death in an alley. There was nothing remarkable in this, of course—such deaths ocurred every night. The deed to his allotment had been recorded. He was no longer available to testify as to the circumstances of the signing. No one could prove any connection between the events.

These tragedies and injustices were occasionally relieved by humorous turns. Once, sensational stories told of the kidnaping of a Creek minor, pictured as a helpless youngster snatched from his parents' arms by unscrupulous men who would probably murder him to get an oil lease on his allotment. The young man was later discovered in London having a wonderful time, his trip financed by an Indian contemporary who had decided to go into the oil business. He had agreed to pay for the trip in return for the young man's lease when he came of age. "If anyone wants to kidnap me on the same terms just send them around and see how quickly I will go with them," the young Indian told the authorities.

As the Cushing boom reached its climax, another sensational giant oil field was discovered in Oklahoma, at Healdton, just north of the Texas border.

Healdton was actually discovered twice. In the late nineteenth century a pioneer wildcatter named Palmer, using the spring-pole method, had kicked a hole down 425 feet and found oil. The joy of discovery was his only reward. These hills were Indian land and as yet there was no legal way to lease them. He told a few friends of his find, swearing them to secrecy, saying that someday when laws would permit, he would open a great oil field there. Then he went to South America in search of other treasures, but found only death.

As a quarter of a century went by, the story of the secret discovery gradually became common knowledge, but nobody particularly believed this typical old prospecting tale.

In 1908, Wirt Franklin, a young Illinois lawyer, moved to Ardmore, the largest town in the area, and looked for a farm as an investment. Impressed with the possibilities of one, he asked the farmer about the water. The farmer took him to the well.

"Now, it's got tar in it, but it's real healthy drinkin' water," the farmer said. "It ain't hurt us none."

He showed Franklin how to strain the "tar" out through cheesecloth. Franklin knew oil when he saw it, and after buying the farm he heard the story about the old plugged Palmer well. He found it on the adjoining property. The ground around the old hole was saturated with oil seeping up from below. Finding other water wells with oil shows, Franklin began leasing the nearby farms. He quickly discovered he had competitors—Roy M. Johnson, editor of the Ardmore *Statesman,* and A. T. McGhee, a carpenter and self-taught geologist. They had found the old Palmer well, too, and were excitedly leasing land.

The men joined forces and together leased about 7,000 acres. Their enthusiasm was such that Johnson mortgaged his newspaper to pay his share of the leasing expenses. The partners soon found themselves in a predicament diametrically opposite to the one that had defeated Palmer. He had drilled a well but could not get leases—they had all the leases but no money to drill a well.

At this point providence appeared as a divine, the Rev. J. M. Critchlow, of Titusville, Pennsylvania. This Scottish minister had been sent to Oklahoma by Scotch financial interests to look for oil prospects. After seeing the Palmer well, Reverend Critchlow looked no further, but agreed to drill three wells for a half-interest in the group's leases.

The first well, Wirt Franklin No. 1, flowed over the derrick for 25 barrels a day; the second well flowed 300 barrels. There was a greater rush to Healdton than there had been to Cushing for this new area was a "poor man's field." The 1,000-foot wells cost only $5,000 to drill as compared to $12,000 for a 2,200 foot well in Cushing. This was the first field discovered in southern Oklahoma, so there was a frenzied trading in leases in the whole area. All those who had not made their fortune at Cushing were certain Healdton was their big chance.

For some it was. One driller parlayed a 20-acre lease into a $340 million oil company. Like thousands of other drillers, W. G. Skelly had left the Pennsylvania Oil Region and followed the booms from Indiana to Illinois, from Texas to Oklahoma. He had managed to buy a drilling rig, and when Wirt Franklin and his group were looking for a contractor to drill ten wells on some of their leases, the contract was given to Skelly, one of the first arrivals after the discovery.

Many a driller has gone broke trying to make the transition from driller to operator, but Skelly bought a 20-acre lease that turned out to be in the most productive part of the new field. With this working capital, he shrewdly continued to follow the booms in Kansas, Oklahoma, and Texas until the Skelly Oil Company became one of the most successful independents in the mid-continent area.

An Ardmore schoolteacher, Miss Odessa Otey, hit a double jackpot. When Roy Johnson and his partners were desperately in need of money to carry out their leasing campaign, she invested her savings with them. She not only helped find a giant oil field but celebrated the discovery by becoming Mrs. Roy Johnson.

The excitement of this carnival of oil could not fail to appeal to the showman in John Ringling when his circus pitched its tents in southern Oklahoma. Several years before, he had become friendly with Jake Hamon, an Ardmore youngster, when Jake spunkily boarded his private car demanding payment of roustabout wages due him. Jake Hamon had been among the first to get leases at Healdton. Needing financing, he turned to Ringling. Ringling bought an interest in his leases and financed the building of a railroad from the Healdton field to Ardmore to haul this great new flood of oil. The Oklahoma town of Ringling sprang up around the railroad terminal, and the Ringling private car was a familiar sight on the siding as the circus owners increased their holdings in the field. Mrs. Ringling bought leases and drilled wells on her own account.

It would not be long before Jake Hamon would have a town named after him, too, the boom town of Jakehamon, Texas. His son, Jake L. Hamon, would become one of the most prominent independent oilmen and a leader in industry affairs.

Healdton's boom community was Ragtown. In a typical ten-day period in Ragtown, two men were held up and shot; a woman was burned to death in a raided whisky joint; a drunken man beat his wife almost to death with his wooden leg; a group of men who had won the pot at a gambling house tried to flee with their winnings, were pursued and beaten to death; two men were shot and two killed in another gambling joint; a man was found dead in a church following a prolonged drunk; another man died who had been drunk for twelve days. To top it off, a bandit accosted a man at night and shot him fatally when he failed to raise both his hands. The deceased had but one arm.

Reverend Critchlow hired another Presbyterian minister, Rev. Albert M. Bean, to be pastor of the oil-field church he established on the group's leases. Services were held each Sunday morning and evening, and attendance was compulsory for all employees of the Dundee Petroleum Company, which Reverend Critchlow and his Scottish associates had formed.

Reverend Bean reported candidly that

"the millenium has not as yet been ushered in to this place. The saddest mission I have to perform is to answer letters from mothers whose boys were killed or had died in drunkenness and rather than tell the facts in the case, I would ask someone else to write, which usually read like this:
DEAR MADAM: We herewith tender you our heartfelt sympathy in the loss of your noble son. He was a bright spot in our circle and we sorely miss him. The floral offerings at his funeral and the excellent casket in which he was buried were the best token of his large circle of warm friends.
I couldn't tear the bleeding heart of the mother open by telling her that her boy was shot or had frozen to death while drunk or had been found dead after a prolonged debauch."

Oklahoma oil men nearly drowned in Cushing and Healdton oil. As the state rocketed to first place, producing a third of the nation's oil, the price plummeted to 40 cents a barrel.

"Stop the drill!" was the oil producers' desperate cry. At a mass meeting Cushing operators, led by Harry Sinclair, agreed to drill no new wildcats, only offsets to wells that would otherwise drain their leases. But of course this remedy resulted in a new round of gushers.

"Barbers in Drumright have organized a union," the *Oil and Gas Journal* jibed. "They say that there will be an immediate increase of prices for a shave if oil producers again wear such long faces as they did during the temporary shutdown. The price of a shave will be governed by the length of the face."

What seemed a catastrophe in 1914 became a blessing in 1915. There was need for every barrel. American consumption doubled between 1912 and 1915. A million cars were sold in 1915, half of them Fords. The industry's assembly lines were preparing to produce 1,600,000 the next year. For the first time cars were sold on the installment plan. With the price of a Model T down to $390, almost everyone could buy a car. Farmers were beginning to buy tractors, too, and there were 300,000 motorboats on American lakes and rivers.

Europe turned to the United States for oil to fuel its war, while the producing nation determinedly maintained its uneasy neutrality.

Harry Sinclair went to London and sized up the situation more realistically. He cabled his partner, Patrick White, that people in England had "settled down to the idea of a long drawn out war." However, he reported happily that financial conditions were "normal" and that he had closed a deal with the Royal Dutch-Shell to pay for half the cost of a pipeline from Cushing to the Gulf Coast. Sinclair could see a short cut to his goal by an alliance with Standard's most powerful competitor.

Royal Dutch-Shell had been organized in 1902 to fight Standard's world monopoly. Dutch and English groups that had developed oil production in the East Indies in the 1890's were unable individually to compete with Standard in world markets; therefore they merged into one company. Finding that Standard continued to control world prices because it monopolized the American market and could charge up foreign losses in an international price war to its domestic business, Royal Dutch-Shell entered America in 1912, determined to compete with Standard on its home ground.

In Oklahoma, Shell formed a company to buy producing properties. The new company's name contrasted strangely with the forthright exuberance of the native names, but there was a point to it. The English wife of the Shell agent selected Roxana, the name of Alexander the Great's wife, which meant in Persian, "Dawn of a New Day."

Roxana bought some choice producing leases in both Cushing and Healdton. The idea of a pipeline to the Gulf Coast to lessen dependence on Standard was as agreeable to Royal Dutch-Shell as to Harry Sinclair, who jubilantly announced that he had purchased a refinery site at Baton Rouge and was buying the right of way for the 365-mile supply line.

Then something incredible happened. Harry Sinclair was accused of being a tool of Standard Oil.

To guard against Standard's monopolistic practices, Congress had passed a law prohibiting anyone from owning an interest in more than 4,800 acres of oil and gas leases on Indian lands. The Department of the Interior interpreted this as meaning that if a single stockholder owned stock in a corporation which owned that amount, it would disqualify any other corporation in which he held stock from owning any Indian leases. Sinclair requested that this interpretation be revoked so that a company could be formed of owners of Cushing production which would have large enough assets to borrow money to finance its half-interest in the pipeline.

The Department of the Interior was swamped with telegrams protesting this request and asserting that Sinclair was in league with Standard Oil. A petition, signed by four hundred Tulsans, was sent to Washington. Sinclair indignantly demanded that the Secretary of the Interior wire the American Ambassador in England asking him to see Sir Marcus Samuel, head of the Royal Dutch-Shell combination. The Secretary replied that even if Sir Marcus would verify Sinclair's statement "it would not be evidence that Standard Oil money was not behind the pipeline." Royal Dutch-Shell was as startled as Sinclair.

"We have 'got the hooks,' " Sinclair proclaimed angrily. "Somebody else can lay the pipelines. We will find a market for our production. Our purpose now is to get rid of our oil at such price as will yield us a profit whatever the posted quotations are. We have been sustaining the market for months and all the thanks we get is a knock."

Defeat meant to Sinclair that there was a better way to accomplish his purpose, and now he had the impetus and inspiration to go ahead.

Although America had been producing oil for over half a century and oil had become the country's most important natural resource, no real producing science had yet been developed. When oil was discovered, no one knew how much was there, what its quality was, or what the most efficient methods of recovering it were. A well would roar in for 5,000 barrels a day and shortly afterward would settle down to 100 barrels or less. No one could tell when it might go dry. Because of this uncertainty, banks refused to make loans, and this was what kept the independent producer in financial bondage. This was why he quickly sold his flush production to a big company for what would often prove to be a ridiculously small sum. This was why an independent could seldom raise enough working capital to become an integrated company with its own pipelines, refineries, and marketing outlets.

Sinclair had already tackled this problem. Unable to get a bank loan on his production, he had established his own bank, the Exchange National Bank of Tulsa. Subsequently the National Bank of Tulsa, it was one of the first great oil banks in the country.

However, Sinclair's own bank was not yet big enough for what he had in mind. One year after the defeat of his pipeline plan Sinclair had put together a $50-million oil company, the fourth largest in America and the largest strictly independent oil company. He had 7,000 barrels of oil production a day, five refineries, 361 miles of pipelines in Kansas and Oklahoma, and was preparing to build a pipeline from Drumright to Chicago. He had risked everything, spared nothing, and when he had finished, men could see the result, although no one could really understand how such a plan could have been conceived and carried out so rapidly.

The man's rise was so meteoric that it was popularly attributed to his sheer luck at finding oil. A story even circulated that he had gotten his start by shooting off his toe to collect insurance in order to finance his first wildcat. When a friend once asked him, years afterward, if this were true, Sinclair laughed and said, "It makes a good story, doesn't it?"

What the public did not know was that Harry Sinclair had set his course from the day almost ten years before when John D. Rockefeller, Jr., had patronizingly commented that he seemed to be doing very well, and he had silently resolved to show the Rockefellers how much better he could do.

The accusation of being a tool of Standard Oil had been the fire that released the quicksilver of action from the cinnabar of resolve. In one brilliant move after another he had dissolved his partnership with Patrick White—there could be but one general in this campaign—and sold their Cushing and other producing properties to the Tidewater Oil Company for $10 million. He had then started drilling some 20,000 acres of unexplored leases purchased from his ex-partner. Before anyone quite realized what he was doing he also concluded a breath-taking buying campaign of producing leases, refineries, and pipelines. Then he planted his flag in Wall Street. Five New York banks underwrote his $50 million Sinclair Oil and Refining Company. Cushing had indeed been a giant step forward.

With the poker face that was one of his greatest assets, this new colossus of oil said: "I want to deny the report that my company is organized to compete with Standard Oil Company. We expect to produce oil, to refine it and to sell it in the markets of this country and the world in competition with anybody."

Only a few appreciated how keenly Harry Sinclair enjoyed announcing that a member of his board of directors was Theodore Roosevelt, Jr.

Edwin L. Drake (in stovepipe hat), soon after the historic
discovery.

The Lucas gusher, Spindletop,
Texas; January, 1901.

Kicking down a hole: the old
spring pole method, still used in
the early 1900's.

Shell Oil Co.

Standard Oil (Ohio)

Benjamin Silliman's
report, with George
Bissell's inscription.

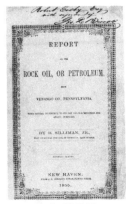

Drake Well Museum

American Petroleum Inst.

"Forty thousand strangers descended on Beaumont." The Spindletop boom, 1901.

Spindletop soon after the big discovery: "You could step from one rig floor to another."

Fred A. Schell

One of Spindletop's fires.

Harry Sinclair (left) and Senator John Overfield, an early associate, soon after Glenn Pool.

Anthony F. Lucas.

Everette Lee DeGolyer (seated); Mexico, 1910.

Difficult transportation and bad roads: typical features of the transient boom towns.

Mrs. E. L. DeGolyer

Standard Oil (N. J.)

After Canfield and Doheny found oil (1893), "West Los Angeles began to look like a gopher city."

Hub of activity for Cushing field: Drumright, Oklahoma, 1914.

John H. Galey.

The industry's first geological team: Cities Service geologists, 1915.

Dad Joiner's No. 3 Daisy Bradford: a laughingstock in the industry, it discovered the nation's greatest field.

Dad Joiner (white shirt and tie) receiving congratulations from geologist A. D. Lloyd.

E. W. Marland.

Carter Oil Co.

Seminole, Oklahoma, 1927: "Cushing and Drumright all over again
—clapboard joints, saloons, tents, mud."

Gusher-drenched crew, Yates field, Texas, 1929.

Ohio Oil Co.

Wild Mary Sudik roaring gas and raining oil, March, 1930. For eleven days the well held Oklahoma City at bay.

Mike Benedum (right) and Levi Smith at Big Lake, Texas, 1924.

A. I. Levorsen.

Wallace Pratt (left) and other Humble geologists on a field trip. Geologist Morgan Davis (third from right) later became president of Humble.

Conroe, Texas, discovered 1931. One of the first fields drilled with well spacing to prevent waste. Compare views of Spindletop and Los Angeles.

BOOK THREE

Minds
of Men

*It is the genius of a people that de-
termines how much oil shall be reduced to possession; the presence
of oil in the earth is not enough.
Gold is where you find it, according to an old adage, but judging
from the record of our experience,
oil must be sought first of all in
our minds*

WALLACE PRATT

1 : *Divining rods and doodlebugs*

IT IS CURIOUS that men did not begin scientifically to study the earth beneath their feet until the eighteenth century. For thousands of years they dug holes and pits in search of rocks, minerals, and water without ever observing that certain rocks always lay in a definite order, each layer revealing a chapter of the earth's history. The stars had more meaning for them than the fossilized remains of plants and animals that studded the rocks, mutely dramatizing the planet's development over 2 billion years. Men found natural resources by chance or else—the same thing—they relied on divining rods.

In the early sixteenth century German miners believed that a forked twig, usually from a hazel bush, would dip over a buried ore vein. The "scientific" idea was that as hazel bushes grew over ore bodies they had an affinity for ore. Variations developed. Hazel twigs would find silver, ash would find copper, pitch pine would find lead and tin, iron and steel rods had an affinity for gold.

Divining spread rapidly throughout Europe. Emperors, dukes, and even popes sanctioned the use of rods to find minerals and buried treasure. Priests and laymen alike used them to find water. Today, divining still flourishes. The United States Geological Survey receives so many inquiries about the art that it answers with a form letter to the effect that divining enjoys no scientific standing. Despite such discouragement 20,000 diviners, principally water witches, were blithely plying their trade throughout America in 1978.

Naturally, when Edwin Drake's well made finding petroleum a lucrative business, the divining rod was suddenly found to have an affinity for oil. The science of geology was then young, and what few geologists there were in America were trying to unravel the basic problems of the origin and age of the earth, how mountains were formed, the effects of wind and water, and the classification of rocks. These things had to be settled first before geology could be economically useful in finding natural resources. When the Pennsylvania oil seekers exhausted the natural signposts of oil seeps, they found divination as reliable as random drilling. Belief in the method increased.

Spirit mediums were also helpful. In the 1860's Mrs. Hall, an Oil City medium, described to Jonathan Watson the exact spot to drill, predicting a 300-barrel flow. The cagey Watson, part owner of the land on which the Drake well was drilled, refused to pay Mrs. Hall her customary $10 fee, promising she could have the first day's production if the well came in. He had a precedent: Cicero, two thousand years before, had praised the sagacity of a man who told a diviner "I will pay you out of the treasures which you enable me to find." However, Mrs. Hall's check was for $1,650 —Watson's well produced 330 barrels the first day instead of 300.

Abram James was a Pennsylvania spiritualist whose guide was Lalah, an Indian maiden dead some three hundred years. One crisp fall day in 1867, James was driving from Pithole City to Titusville with three oilmen. A mile from Pleasantville, Lalah assumed control of James, flinging him over a fence and throwing him violently on the ground, where, before losing consciousness, he dug his finger into the dirt. When his astonished companions revived him, he said that Lalah had revealed that he should drill at the spot his finger had marked.

The oilmen did not share James's confidence in Lalah and refused to drill. James went ahead on his own, leasing the farm and asking the Titusville Second National Bank for a drilling loan. When the bank's president refused, James informed him that Lalah herself had said the Second National Bank would loan the money. Lalah was right on all counts. The bank loaned the money and James opened up the Pleasantville oil field, finding oil at 850 feet and drilling twelve producing oil wells.

Those believing neither in divining rods nor spiritualists could choose from a fascinating variety of other methods. There were oil "smellers." An Ohio man claimed he could locate gas fields by the tension in his neck muscles. Another man, dangling a bottle of crude oil on a string, declared that he received electrical shocks whenever he walked over oil.

As discoveries increased in importance, the simple divining rod evolved into remarkable machines, often dubbed doodlebugs, each with its own believers. A Pittsburgh scientist announced a giant X-ray machine which he would focus on the earth and, by electrical current, excite its molecules, separating them and making the earth temporarily transparent so that he could see where oil was.

In 1913 the Little Giant Oil Company was organized in Beaumont to develop "a great belt of oil" that an equally gifted machine had located near Spindletop. This doodlebug had been invented by a German scientist, who one day was found dead in his laboratory, the instrument in his lap, a peaceful smile illuminating his face. His son concluded he had died of joy upon perfecting the machine. Oilmen began believing in the machine when they saw it locate a gold watch and silver dollars. It did much less well, of course, when it came to finding production for the Little Giant Oil Company.

Most oilmen classified geologists with diviners and doodlebuggers. As a matter of fact, the man with a machine was more apt to obtain a hearing than the man with an idea. The doodlebugger expressed no doubts as to his ability to find oil. The geologist qualified every belief with "perhaps," "might," or "it is logical to believe."

Oilmen had been finding new pools in increasing numbers for fifty-four years by following seeps, trends, and hunches, when Charles N. Gould, founder of the Oklahoma Geological Survey, presented a paper at the International Geological Congress, in Toronto, in 1913, summarizing the latest beliefs of petroleum geologists. His cautious statement that "careful studies of geologic conditions have demonstrated that there is a rather definite relation between structure of rocks and occurrence of oil and gas" could scarcely be expected to fire oilmen with enthusiasm for the new science or a desire to risk their drilling money on its recommendations.

Gould barely kept his scientific audience awake with his statements that all the big pools of Oklahoma lay under anticline folds and that if well-marked anticlines were drilled they would probably yield considerable amounts of petroleum. Geologists had been saying much the same since the days of the Drake well, but almost none of the oil fields in the old oil regions had been found in anticlines. Why should an unsuccessful guide to prospecting in the East be successful in the West?

Most oilmen, as a matter of fact, hardly knew what *anticline* meant. However, it would soon become the most important word in the exploration vocabulary, and the search for oil would become a search for anticlines—the places where rock strata arched, dipping or inclining in opposite directions from a ridge, like the roof of a house. Gould's paper mentioned that the Cushing field was an anticline. The following year, the significance of his statement became clear when the Oklahoma Geological Survey published a structure map of Cushing's surface geology. No one needed to be a geologist to understand the picture—one of the world's greatest oil fields was coming from a big structure whose arching could easily be seen and mapped.

And Charles Gould had said that other such formations would probably yield petroleum. Where were these anticlines? The rush to find them began, and the majority of discoveries during the next fifteen years resulted from the industry's sudden acceptance of petroleum geology as a science. Cushing changed the course of oil history as spectacularly as Spindletop had done. Most of the large companies established permanent geological departments. Universities began to offer four-year petroleum-geology courses. If an oilman had a geologist's favorable report on a prospect, he could be sure of finding someone ready to risk money on it.

There had been, of course, some earlier scientific successes. In 1897 California's peculiar oil-finding problems led the Southern Pacific companies to hire Edwin T. Dumble, a Texas state geologist. The following year Lyman Stewart hired W. W. Orcutt to found Union Oil Company's geological department. However, California was so remote from the rest of the oil scene that developments there had no influence in proving petroleum geology's value elsewhere.

Lord Cowdray credited his success in Mexico in 1910 to two American geologists—Dr. C. Willard Hays, former chief of the United States Geological Survey, and the brilliant young Everette Lee DeGolyer. Royal Dutch-Shell was using geologists to explore in the East Indies.

In the mid-continent, where petroleum geology was finally accepted in 1913–1914, Gulf Oil had led the way. The Mellons hired M. J. Munn of the United States Geological Survey in 1911 to form the first geological department of any oil company in the region. After Cushing's discovery, Munn immediately mapped the anticline, enabling Gulf to buy choice leases that ultimately produced more than 11 million barrels of oil.

When the petroleum geologists of the Southwest held their first meeting in January, 1916, to form an association, there was no discussion of how geology had found oil in America. Instead they listened to DeGolyer tell about the geology of the Mexican oil fields.

Although they had not yet proved they could find oil, these young scientists discussed the advisability of licensing geologists to distinguish them from the quacks and doodlebuggers. But licenses would not be needed. The geologist's spectacular achievements would soon be proof enough. DeGolyer himself would soon introduce the most important of oil-finding tools. A timid young Bureau of Mines geologist, Wallace Pratt, just returned from the Philippines, read a paper on a new technique of geological mapping; he would do more than any man to make the public aware of the importance of geology in the art of oil prospecting. A young Texan, William E. Wrather, who spoke on West Texas oil possibilities, would soon start a great boom there.

Charles N. Gould read no paper but listened with pride to these younger men, most of whom had been his students at the University of Oklahoma. Although it would be several months before anyone at this meeting would realize it, Gould had already brought about the first major achievement of petroleum geology in America—the discovery of two giant oil fields in Kansas.

As early in 1912 this modest geology professor posed the question "Why bother with a geologist at all?" Addressing the Natural Gas Association of America, Gould said, "If a geologist is honest, if he has training and experience, if he has studied a region not casually, but carefully and painstakingly, sometimes for months and years, he will possibly be able to eliminate part of the risk."

Such candor was not calculated to persuade his listeners to bother with geologists. Gould should have mentioned that by just such careful, painstaking study the year before he had mapped an Oklahoma anticline, discovering a new gas field.

Nevertheless, Gould won a convert. A. J. Deischer was manager of the Empire Gas and Fuel Company, one of several companies owned by Henry L. Doherty; his problem was to find new gas reserves to supply various Kansas towns and industries. Deischer persuaded Doherty to hire the geology professor.

A believer in the value of scientific knowledge, a profound thinker, a man who never stopped studying, Henry L. Doherty was a self-educated engineering genius who eventually obtained 140 patents covering his inventions and processes in the gas, cement, chemical, coke, smelting, petroleum, and house-heating industries. Leaving school at age twelve and starting as an office boy in a Columbus, Ohio, gas company, he went on to build a great industrial empire.

While young Doherty was poring over gas-machinery catalogues, teaching himself how a gas plant worked, another Ohio boy, Charles Gould, two years older, was working on his parents' farm near Marietta and studying at night to be a schoolteacher. When the Gould family migrated to Kansas, eighteen-year-old Charles taught country school. He managed to get a degree from Southwest Kansas College after seven years divided between study and work.

The world of science opened up for him when he chanced to find some fossil shells. A paleontology professor at the University of Kansas, to whom he sent the shells, aroused his interest in geology. Charles Gould then set himself three goals. He would learn everything possible about the geology of the southern Great Plains, he would become head of a state university's geology department, and he would become a state geologist.

In 1900, after Gould received his master's degree for his study of fossils, he was invited to establish a geology department at the University of Oklahoma. He had accomplished one goal. The next was achieved when Gould succeeded in having Oklahoma's new state constitution provide for a State Geological Survey. Continuing his university work, he became the Survey's director in 1908. After the discovery of Glenn Pool in 1905, Gould expanded his last goal of knowing everything possible about the regional geology to include doing everything possible to prove the usefulness of geology in finding oil.

This shy, kindly man was tireless in his efforts. There were no books to go by. He and his students had to go into the field and discover the facts from which books could be written. Sturdy legs were as important as curious minds in Charles Gould's classroom, which was the earth.

Working from the known to the unknown, Gould developed his maxim: "Any anticline is worth drilling."

No one knew then, nor do they now, just how oil and gas originated, moved in the earth, and accumulated in pools. However, Gould and his students carefully studied the ages and types of porous rocks that were favorable for trapping oil. They found where these rocks outcropped and measured the dip of the strata over a wide area to determine if the rock layers had been tilted to form a trap where oil might have collected. They began to accumulate data that would prove there were other types of traps besides anticlines. This would explain why the oil in the old Eastern Oil Region had not been found in anticlines.

Feeling that his university's geology department and the State Geological Survey were well established, Gould resigned from both in 1911 to work as a private consultant. Two years later he was in Kansas, helping Henry L. Doherty's company look for new gas supplies. Some small gas wells had been found near Augusta, Kansas, in the southern part of the state, as early as 1906. When Gould examined the area, he found they were located on an anticline. A river had cut away most of the rocks that would have made the structure obvious, but the remaining clues were enough for Gould.

The anticline extended for 12 miles. Cities Service Company, as Doherty's organization would soon be called, immediately leased some 10,000 acres along it. Gould recommended that more detailed work be done to select the most likely location for a well, and suggested the hiring of his nephew, Everett Carpenter, a junior geologist with the United States Geological Survey. The first well, drilled in 1913, opened a new gas field. The following year when a development well failed to find gas at 1,400 feet, where other wells had found it, Cities Service drilled deeper and found some small production in what would prove to be a 51-million-barrel oil field.

The importance of this geological success went unnoticed. Augusta was not even a sideshow to the big circus going on at Cushing and Healdton. However, Cities Service encouraged Everett Carpenter to build a geological staff and conduct a survey of a wide area in search of other favorable places.

The little town of El Dorado lies about 17 miles north of Augusta. In 1909, following the discovery of the small, shallow gas wells at Augusta, the three thousand citizens of El Dorado voted a $20,000 bond issue to wildcat for gas and oil, hoping to develop a community supply. They sought the advice of Erasmus Haworth, a University of Kansas geology professor, who found an anticline near the town and recommended drilling a well on its crest. The well was a dry hole, as was a second drilled in the town park.

After the 1914 Augusta discovery, El Dorado's town council elatedly announced there was enough bond money left to drill another well. They drilled 3,000 feet, twice as deep as their other two tries, but with no more success.

While they were drilling, Everett Carpenter and his rapidly expanding staff of geologists could scarcely believe the results of their field studies. Unaware of Haworth's work, they traced the El Dorado anticline and discovered that it covered 34 square miles of what they believed was favorable oil land. This was more territory than the great Cushing field covered. Was it possible they had found another giant?

Cities Service now had such confidence in geology that company officials paid no attention to El Dorado's three dry holes. They leased 30,000 acres, including the town's 790 acres of leases, reimbursing the town for the entire expense of its last dry hole. In addition, Cities Service agreed that if a test well on its own leases did not find gas or oil, it would drill another test on the town leases.

Early one September morning in 1915 the test well started down. Carpenter and the other geologists who had mapped the structure began to feel the tension of the poker player who suddenly realized that his final bet is beyond his pocket and is sure his hole card isn't a winner. This was the first time in history that a company had conducted a wide-area geological survey and unhesitatingly leased the vast amount of acreage its geologists recommended. If it was not productive, the reaction against geology would be in proportion to the ambitiousness of the project. This well would make history one way or another.

At only 660 feet the well started to flow. Almost the entire population of El Dorado hastened jubilantly to the scene. A whole town had experienced a wildcatter's emotions from the disappointment of dry holes to the ecstasy of victory. The original objective of a municipal gas supply was forgotten in what Mike Benedum described as "awe at the realization that you have triumphed over a stubborn and unyielding Nature, forcing her to give up some of her treasure."

The well wasn't a gusher but excitement was as great as if everybody knew what the deepening of the well and other wells would soon reveal—that this was a giant oil field, the greatest ever discovered in Kansas, and one of America's great ones. Thus, the first concerted geological effort found 275 million barrels of oil. Furthermore, of the 34 square miles Cities Service had leased, all but 7 square miles would

prove productive. Only 3½ square miles of the field lay outside the geologists' recommended area. Never before had
one company owned almost all of a new discovery—never
before had there been a way to judge the extent of an oil
pool in advance of drilling.

The dry holes on top of the anticline had also proved that
dry holes did not necessarily mean the absence of oil. The
rocks might not be porous enough at the top of a structure
to hold the oil.

Because Cities Service had leased practically all the land,
the drama of Cushing was not re-enacted in Kansas. The
boom this time was in men's minds instead of on the land.

Doherty directed Carpenter to explore all the mid-
continent states; for the task, Carpenter assembled a staff of
250 geologists. Their work would lead Cities Service to even
more sensational discoveries, eventually making it one of
the nation's ten biggest oil companies.

Charles Gould, the godfather of this great geological
victory, was beseiged by clients.

One afternoon he reluctantly handed a report to two men
from Amarillo, Texas. M. C. Nobles, a wholesale grocer, and
T. J. Moore, a traveling salesman, had caught the fever
and leased some land in southern Oklahoma and asked
Gould to examine it.

"I'm sorry," he said, "but I can find no geological reason
to recommend your acreage."

The disappointed Nobles, about to take his leave, chanced
to ask if Gould knew of any prospects near Amarillo.

"No, I don't," Gould replied. No one had ever looked for
oil in the great prairies, mesas, and buttes of the Texas Pan-

handle—country unlike any in which oil had been found. Suddenly, Gould recalled that, eleven years before, he had mapped three anticlines near Amarillo. How could he, who taught that "any anticline is worth drilling," have forgotten them? With chagrin, he realized how compartmentalized the scientific mind can be at times. He had been looking for water, not oil, that summer when he mapped the area for the United States Geological Survey.

As he told Nobles and Moore about the anticlines, all that he had learned about oil and gas began to relate itself to this unexplored terrain. The Texans had no difficulty in persuading Gould to return to Amarillo with them.

Gould mapped the anticlines in detail, and Nobles and Moore leased some 70,000 acres covering them. To them it seemed almost too great a dream, for this was more than twice the acreage Cities Service had leased at El Dorado. However, the Texans were supremely confident. They even laughed, as though he were joking, when he told them, as they watched the first well start, that he couldn't guarantee there would be oil there.

"This is the most favorable location for oil and gas," he said, more to reassure himself than Nobles and Moore, "and I think it is more likely to be gas than oil."

When 10 million cubic feet of gas roared from the hole on December 13, 1918, not even Charles Gould realized what a prophet he was. Twenty years would pass before he was given full credit for locating the discovery well for the world's greatest gas field—one vast L-shaped structure extending from the Texas Panhandle across the Oklahoma Panhandle and into Kansas. Along its northern flank in Texas lay four giant oil pools containing more than 600 million barrels of oil. Instead of measuring 70,000 acres, this empire of gas and oil covered 5 million productive acres.

It was enough for everybody at the moment that a promising new producing area had been found, that Amarillo had a new gas supply, and that geology had dramatically scored once again.

The principal skeptics, after these demonstrations of applied geology, were Standard Oil and Harry Sinclair.

Prairie Oil and Gas Company, the mid-continent company that had been Standard's biggest producing subsidiary and, since the dissolution, still sold practically all its oil to Jersey Standard, the decapitated head of the Trust, was the biggest purchaser of crude in Oklahoma and Kansas. Charles Taylor, Gould's successor at the University of Oklahoma, was pleased when his 47 petroleum geology graduates in 1914 were immediately hired by oil companies, but perturbed that Prairie did not have a geologist. He queried Prairie officials about their attitude. "Perhaps sometime in the future geology may be taken more seriously than it is now," they told him.

Among the companies Sinclair had bought in the overnight formation of his new empire was Chanute Refining Company. Primarily interested in its refinery and producing properties, he paid scant attention to the untested acreage that Chanute had leased some 65 miles northwest of Cushing in the wildest of wildcat territory. The Chanute officials had been so intrigued with the possibilities of geology that they had bought the leases from Bert Garber, who had hired a California geologist, Dorsey Hager, to map the area. Chanute then hired a geologist of its own, and Harry Sinclair inherited him. Having satisfied himself that the man "couldn't possibly contribute to the advancement of the company," Sinclair fired him.

As for the untested acreage, Sinclair finally got around to drilling it in 1917—not because a geologist had recommended it, but simply because the company owned the leases. The test well, flowing 200 barrels a day, discovered the 67 million barrel Garber field. Sinclair was elated, but still unconverted to geology—even when a deeper test brought in the biggest gusher Oklahoma had yet seen— 27,000 barrels a day.

Regardless of visions of abundant new oil resources yet to be found, the world as a whole was gripped with the fear of an oil shortage in 1917 when America entered the war.

The use of oil had revolutionized warfare as drastically as the invention of gunpowder. This was a war of fast ships, planes, tanks, trucks, tractors, cars, ambulances, and even taxicabs, a fleet of which had saved Paris in 1914 by rushing troops and supplies to repel the invaders.

Verdun was saved in 1916 by 40,000 motor trucks and tractors which transported troops and munitions to the fortresses. Motorized equipment was the deciding factor in winning the Battle of the Somme in 1916 and the Battle of the Aisne in 1917.

Oil played a decisive role on the seas. The Allied supply line was almost severed in 1917 by German sinkings of Allied merchant marine ships. Winston Churchill called this "the gravest peril which we faced in all the ups and downs of that war." The United States Navy's quick building of oil-burning submarine chasers, submarines, destroyers, and other vessels, with quick turnabout time, greater speed, maximum cargo

space, and smaller crews, won the race against German sinkings.

Even before its entry, America was supplying 90 per cent
of the great flow of oil which was the margin of victory.
Storage tanks were drained and wells opened wide to ship
100,000 barrels a day to Europe. Cushing, Healdton, and
the newly discovered El Dorado supplied most of it. Even
so, there was not enough. "Gasless Sundays" were decreed.

The end of the war increased rather than diminished the
need for new oil discoveries. In 1914 there were only
1,600,000 passenger cars in America. By 1918 there were
5,600,000. Farm tractors had increased from 17,000 in 1914
to 44,000 in 1918. Motor trucks had jumped from 85,000 in
1914 to 525,000 in 1918. The end of the war marked the real
beginning of the aviation industry and the powering of all
the world's ships by oil.

Fears of an oil shortage were intensified. It would take the
entire contents of three giant oil fields, each containing 100
million barrels, and one field containing 50 million barrels
just to replace the oil the machines consumed in 1918. But in
1918 there was no way to judge what the ultimate recovery
of any oil field would be. Geologists and engineers had not
yet learned enough about the behavior and occurrence of
oil and gas to estimate how much might be in place. Nor
had they learned how to recover the maximum amount
possible from a reservoir.

The explorers had discovered more billions of barrels than
they knew, but what counted was how much oil they could
produce daily, not ultimately. Engineers would eventually
know that Cushing contained a possible total recovery of
455 million barrels and that it would take more than sixty
years from the date of its discovery to produce it. What mat-

tered in 1918 was that Cushing, which had once produced 300,000 barrels a day, was now producing only 57,500 barrels a day, and no one knew how soon the wells would go dry.

The other giants were declining as rapidly. Time would prove them to be reliable old workhorses, but this new economic horse race could only be won by the flush performance of new discoveries.

Oil explorers and petroleum geologists faced their greatest challenge.

2 : *"A wildcatter can't quit"*

MIKE BENEDUM and Joe Trees stared with awe at great jungle trees whose trunks were coated with oil higher than their heads. Vines and bushes dripped oil. Streams of oil flowed at their feet.

"Look at that pool of oil over there, Joe," Mike said incredulously. "If you had your hip boots on you could wade in it. I thought we had seen seeps in Mexico, but I don't think there's anything in the world to compare to this."

The little group of explorers that had come to this remote part of Colombia was an odd one. The emotional reactions were as varied as the backgrounds and objectives of the members. Claude Benedum and Graham Trees shared their fathers' enthusiasm, but it was danger that was their tonic rather than the sight of seeps. From the time they had transferred from a river boat on the Magdalena to native canoes on a small tributary, there had been the breathless knowledge that at any moment they might be the target for a flight of poisoned arrows. This was the domain of Motilone Indians, whose other protection against civilization was a jungle as savage and mysterious as themselves.

It had been great sport for Claude and Graham to shoot at alligators slithering into the water at their approach, and to guess at the length of giant boa constrictors whose coils could be glimpsed in the trees. The young men were unmindful of the soggy, debilitating heat and the swarms of fever-laden mosquitoes. They reveled in the jungle's sense of timelessness. The thought of armies battling at that very moment in France seemed as unrelated to the present as the Gallic Wars. Perhaps this moment of high adventure in Colombia in 1916 belonged to young Benedum and Trees more than to the rest of the party, for within two years they would both be casualties of the war that now seemed so remote.

To Roberto DeMares, the presence of the group at this particular spot was desperately important. For eleven years he had been trying to get someone to develop the million-acre concession he had obtained from the Colombian government. No one doubted the presence of oil here, but the awesome problems of jungle, mountain chains, incredible rainfalls, tropical sicknesses, hostile Indians, no labor supply, and vast distances from sources of supply or market made the task of recovery seem impossible.

DeMares had convinced Luciano Restrepo, a Colombian geologist, and John Leonard, the Pennsylvanian who had obtained the concession in Mexico for Benedum and Trees, that the reward for solving the physical problems would more than compensate the effort and risk involved. These men, in turn, had persuaded Mike Benedum and Joe Trees to see for themselves.

The negative comments of two other members of the party perturbed DeMares. John Harrington, a Standard Oil of New Jersey geological scout, and Newton Graham, a Pennsylvania banker and oilman, talked only of the economic difficulties involved. DeMares hoped that their opinions would not be the decisive factor.

George Crawford, head of a Pittsburgh gas company, and his partner, Milo Treat, were the members of the party with the least to say.

As usual, Mike Benedum got directly to the point. "Well, gentlemen, there's no question but what there's oil here. This is another Mexico, if not bigger. What do you think?"

Newton Graham, the banker, answered first. "I don't want any part of it, Mike. You would have to spend too much money to get it out. After seeing this, it's no wonder nobody has touched it."

"I agree with Mr. Graham," the Standard geological scout said. "I'm afraid I couldn't possibly recommend it to my company."

"What about you, George?"

"Mike, I realize that Mr. Graham and Mr. Harrington certainly are aware of a lot of things about financing and world oil marketing that I don't know anything about," Crawford replied slowly. "However, I've been thinking about the time you and Joe asked me to go to Illinois with you to look over the prospects there. I gave you my expert opinion —that there wasn't any possibility of finding enough gas or oil to make it practical. Look how wrong you and Joe proved me to be. Then Milo and I went along with you on your Mexican venture and we made money on it. If you want to gamble on Colombia that's good enough for Milo and me. We'll go along."

Milo Treat nodded agreement. DeMares began to feel new hope.

"Well, Joe, I guess that's it." Mike smiled. "George, I'm glad you want to come along. Nothing could keep Joe and me out of Colombia after what we've seen. Now, Mr. DeMares, I think we'd better make a deal."

In spite of the magnitude of this new venture, it took only two months to complete the necessary legal work, form the Tropical Oil Company, and sell a million dollars' worth of its shares to a public so eager to invest in a Benedum-and-Trees venture that the offering was oversubscribed five times. Preparing for the actual drilling took longer—two years.

Meantime, back in the United States, there were other projects. Early in 1918 a geologist from Wichita Falls, Texas, entered their Pittsburgh office. There was nothing about William E. Wrather to remind the partners of that resplendent, unknown geologist in Prince Albert coat and high silk hat whose ideas had encouraged them to take their momentous Illinois gamble thirteen years before. This enthusiastic, tall, rangy young man, as much a practical product of the oil fields as Benedum and Trees, had worked his way up through the ranks of the J. M. Guffey Petroleum Company. Convinced that geology was the key to discovery, he had gone out on his own to pit his keen scientific mind against the mystery of the rocks.

Benedum and Trees listened intently as Wrather described how, equipped with hammer, camera, and compass, he had walked hundreds of miles examining the rock outcrops in north central and West Texas. He showed the partners his carefully plotted maps and magnificent photographs. As early as 1902 oil had been found in north Texas along the Red River, separating Texas from Oklahoma. The shallow field gave birth to the town of Petrolia. About 30 miles west of Petrolia, a rancher, W. T. Waggoner, desperately needing water for his cattle, drilled a series of wells as deep as 1,700 feet in search of it. He was infuriated when oil came up.

"I wanted water," he said, "and they got me oil. I tell you, I was mad, mad clean through. We needed water to drink ourselves and for our cattle to drink. I said, 'Damn the oil, I want water.'"

When news of the disgruntled rancher's predicament reached The Texas Company, he was persuaded to lease most of his huge ranch. After drilling several wells, the company found the source of his trouble—a rich oil field that would eventually prove to be another giant. Somewhat pacified, Waggoner insisted on naming it Electra in honor of his daughter.

This oil on the north Texas border appeared to be part of the Oklahoma discovery pattern; it did not stimulate any persistent search further south. But it was the area 130 miles directly south of Electra that attracted Wrather. A few wells drilled in the general area had produced small amounts of oil, but no major field had been found.

As Bill Wrather trudged through mesquite and oak along the banks of Hog Creek studying outcrops, he realized he had found a big anticline. Leasing the land was no problem. On October 1, 1917, all the farmers of the area met in the Hogtown schoolhouse and formed the Hog Creek Oil Company, giving Wrather and his group leases on 6,000 acres for $2 an acre and an agreement to drill a 3,500-foot test well.

This was the farmers' second oil company. Three years previously, hoping that there was oil under their farms, they had subscribed the capital to drill on the village barber's land. At 1,500 feet they ran out of funds. A severe drought then struck the area and farm after farm was abandoned.

Even as the anxious Hogtown farmers were signing their leases, a wildcat well was drilling some 13 miles northwest near the town of Ranger, the result of another emergency mass meeting of drought-stricken farmers and businessmen. W. K. Gordon, general manager of the Texas Pacific Coal Company, which operated near Ranger, had persuaded the New York owners to let him drill a few shallow wells. When a committee from Ranger approached him with the idea of drilling a deep test, Gordon was able to overcome his company's fixation on coal; the company agreed to drill a 3,500-foot test in return for a big block of leases. The Ranger owners came through with leases on 25,000 acres.

While the Hogtown farmers were congratulating themselves on the bright prospects of the well Wrather and his group had promised to drill, the Ranger community was enveloped in gloom. The Texas Pacific Coal Company's first wildcat found gas, but at 3,400 feet the well was abandoned when a drilling bit broke off in the hole. The second well was drilling at 3,200 feet without a show of oil when orders came from the New York office to abandon it. Fortunately, W. K. Gordon insisted on continuing. On October 21, 1917, 231 feet deeper, the drought was ended by a rain of oil flowing over the land at the rate of 1,700 barrels a day.

There was only one dissatisfied person in north central Texas that day. Mrs. J. H. McCleskey, on whose farm the discovery well was drilled, was furious at the sight of her flock of beautiful white leghorn chickens covered with black oil.

Hogtown was almost as jubilant as Ranger, certain that it, too, had harvested its last crops of pigs, peanuts, and poverty.

Both the towns were perplexed when the new discovery did not result in an immediate sensational boom. Despite the nation's oil shortage and the seeming proof that here was a new, abundant source, speculators and wildcatters did not pour into the desolate little towns. What the farmers did not realize was that oilmen had learned from bitter experience that they could lose money as well as make it by acting too quickly. The fact that the Texas Pacific Coal Company held a 25,000-acre block might mean that they owned all the oil field, but until other wells were drilled, who could tell how big or how rich the field might be? And the discovery hardly proved the existence of oil 13 miles away at Hogtown.

On January 1, 1918, the Texas Pacific Coal Company's abandoned first well, which had been flowing a little gas from its open hole, suddenly shook the countryside, erupting with a mighty flow of gas. Within a few weeks this turned into a great gush of oil.

But not even Ranger's second well started the boom. Wrather's group had exhausted its capital in acquiring leases and could find no one in Texas with confidence enough in geology to believe that Wrather's anticline contained oil. At this point Bill Wrather went to confer with Benedum and Trees.

As Benedum listened to the story, he was convinced this fledgling science made sense. It fitted all he had learned in twenty-two years of hunting oil. Besides, it would be ridiculous not to believe that the occurrence of oil in the earth followed consistent scientific laws and principles which could be learned by study. Science was daily proving the obedience of all matter to immutable universal laws. Why should oil be an exception?

Here was the kind of opportunity Mike Benedum liked: a chance to prove a new idea in a new area where others were reluctant to gamble. There were no seeps to encourage him, but what were the Ranger discoveries if not man-made "seeps" to prove that this was real oil country?

"I think we ought to drill this one, Joe," Benedum said.

"I'm sorry, but I can't see it," Trees replied. "Everything Mr. Wrather says sounds convincing, but there has been considerable drilling in this part of the country and no big fields have been found. This Ranger discovery might be just a flash in the pan. You know there have been lots of discoveries like it in West Virginia that turned out to be nothing at all, and when we stepped out a few miles we only got dry holes. You'd better count me out this round."

Benedum was startled. At the start of their partnership they had agreed that they would be free to disagree, but they had reacted alike to every project for so long it seemed incredible that they were at opposite poles on Wrather's prospect.

"It won't feel right, Joe, but I'm going ahead anyway," Benedum concluded.

When the Texas Pacific Coal Company's third and fourth wells came in as gushers in March, 1918, Trees simply smiled at Mike's "Well, what do you think now?"

Joe Trees was the only oilman in the country who wasn't convinced. A frantic drilling campaign spread out like the spokes of a wheel from the hub of Ranger. When Trees pointed out that all the wells drilled east and south, in the direction of Hogtown, were dry holes, it was Benedum's turn to smile.

Nobody but the farmers paid much attention to the Hogtown wildcat as it drilled ahead during the scorching summer months. On the night of September 2, as the tool dresser was hammering a bit into shape on his red-hot forge, the well suddenly blew out. A spark from the forge ignited the column of gas into a flaming torch to spread the good news to the whole countryside. Three days later, after the fire was extinguished, the well began producing 2,000 barrels of oil a day and Hogtown—or Desdemona, as its citizens hastily renamed it in honor of the justice of the peace's pretty daughter—soon became as wild a boom town as Ranger.

There was a special madness about the Ranger and Desdemona booms. All the familiar ingredients were there—saloons, brothels, gambling joints. There was even the mud, for with the oil came a deluge of rain. There were spectacular gushers coming in daily to push the frenzied trading to new heights; there were the village populations suddenly swelling to 16,000 and 30,000 people. But this time there was something more. The war had just ended and Americans were weary, confused, and disillusioned. There was a revulsion against morality and idealism. No one wanted to be his brother's keeper any longer. It was a time of "me first," bathtub gin, flaming youth. To thousands of restless, bitter young men discharged from the armed services Ranger and Desdemona were a glittering answer to the question of where to go and how to catch up in a hurry. They poured into north central Texas from every state, bringing with them a brutality generated by the war.

In no other boom did violence develop into mob action,

as here. When a Desdemona department store owner refused to cash field workers' checks, they organized a gang to wreck and loot the store. Next day all reported to work in liberated $25 silk shirts.

When a drunken driller ordered a $1.25 meal in a Greek restaurant and had only a dollar to pay for it, the owner called the law. In the ensuing fight, the driller lost an eye. That night a thousand field workers came to town with sledge hammers and smashed the marble table tops of every Greek restaurant.

Mobs ran law enforcement officers out of Desdemona three times. Professional crime flourished under these conditions as in no other boom. In Ranger, rounding up the night's corpses became a routine affair. Highjackings on the Desdemona-Ranger road were too common to mention.

Eventually, Ranger's indignant citizens solved their own problems. With Rotary Club members acting as leaders, they began raiding and closing. The sight of thousands of dollars' worth of smashed gambling paraphernalia and whisky poured into the streets convinced the lawless elements that it was time to move on. From then on, law was in the saddle.

The stories of sudden riches had a new appeal for a nation eager to re-embrace a philosophy of materialism. The boom became fashionable. It attracted celebrities in numbers.

Tex Rickard, the boxing promoter, drilled some successful wells. Jess Willard, the heavyweight boxing champion, came looking for an investment. Ex-president William Howard Taft delivered a lecture in Ranger, and Billy Sunday held a great revival meeting in the main street. Rex Beach came to collect material for a novel of the oil fields, *Flowing Gold*. The boom gave Conrad Hilton his start: in nearby Cisco he bought his first hotel, calling it "a cross between a flophouse and a gold mine."

The John Ringlings' private car was seen in Ranger as often as it used to be in Healdton. Jake Hamon built another railroad to relieve the incredible congestion of supplies, and the town of Jakehamon mushroomed into existence.

Money changed hands quickly and in fantastic sums. Oil was selling for $4.25 a barrel. A typical gusher would yield its owner a million dollars a month. Two oilmen refused $7 million for a 160-acre lease. One farmer delightedly sold his farm outright for $250,000, taking a job for $5 a day as helper on the wells drilled on it, saying he could save his principal and live on his salary.

A speculator arriving with no capital could make as much as $50,000 profit in a day. The trick was to buy a lease, pay for it by check, and sell it so quickly that he could deposit the funds to cover the check before the bank opened the next morning.

When it was discovered that part of the Desdemona field lay under the town, a well was drilled in the middle of Main Street, merchants and citizens gleefully tearing down stores and houses to make room for more derricks.

The Merriman Baptist Cemetery became famous when, as the Ranger field surrounded it, the church was offered $100,000 for a lease. Will Ferrell's poetic rendition of its refusal was published throughout the country.

> All of oildom knows the answer
>   When the chairman shook his head
> Pointing past the men of millions
>   At the city of the dead . . .

> "Why disturb the weary tenants
> In yon narrow strip of sod?
> 'Tis not ours, but theirs the title,
> Vested by the will of God.

> "We the board have talked it over,
> Pro and con, without avail,
> We reject your hundred thousand—
> Merriman is not for sale."

With a poet's license, Will Ferrell overlooked the fact that title to the cemetery was vested in its "weary tenants" not by the will of God but by the will of a deceased church member, so there was nobody available who could legally sign a lease.

Even seasoned oilmen became overenthusiastic in this hysteria, buying acreage at prices that would have appalled them in other booms. They would soon regret the costly drilling campaign—Mike Benedum most of all. At the moment, Benedum was exultant. Just before the Desdemona wildcat came in, he had received good news from Colombia. After two years the jungles had fulfilled their promise. The oil that had been waiting for so many millions of years seemed impatient to be found. At 80 feet, the first well started to flow 50 barrels a day. When it was deepened to 2,260 feet it gushed 5,000 barrels. Two more wells flowed even more prolifically. The tremendous financial and physical problems of constructing a pipeline and refinery remained, but Benedum was confident they would be solved.

As the wells in the Desdemona field poured out a seemingly inexhaustible flood of oil, Mike Benedum began to think of a new and, for him, strange objective. No other wildcatters had demonstrated oil-finding ability so dramatically or so consistently. He and Trees had always been content to sell their discoveries and go looking for the next great gamble. Why not keep this rich new Texas field, add to it all their other oil interests in America, and form a company which would transport, refine, and market its own discoveries? Like Harry Sinclair, Mike Benedum began to dream of conquest.

Joe Trees was not interested.

"You're getting old, Joe," Mike chided.

Bill Wrather and his associates preferred to sell their interests to Benedum for cash rather than merge with the new Transcontinental Oil Company he was determined to form. It took $23 million to buy the various properties for Transcontinental. New York and Philadelphia bankers provided $20 million and Benedum wrote a check for the balance.

On August 1, 1919, when the New York Curb Exchange market opened, Transcontinental stock was listed at 48¼. Two months later it was being traded for 62⅝. On paper Mike Benedum's stock interest had made a profit of $10 million dollars. The stock seemed sure to spiral to 100. Benedum began to wonder why he hadn't thought of such a company before. A month later, time seemed to have run backward, and he was once again living the nightmare of 1904 when the First Citizens Bank of Cameron, West Virginia, closed its doors.

The great Desdemona and Ranger fields suddenly ceased to flow, and Transcontinental stock crashed to $6 a share. Desdemona proved to be the flash in the pan Joe Trees had indicated it would be. Diplomatically, he said nothing.

The rapidity with which Desdemona and Ranger fizzled was startling. A 7,000-barrel-a-day gusher went dry in five months. Few survived a year. Nature had played a monstrous joke on the oil finders. Just as they thought they were beginning to learn something about the behavior of oil in the earth, their theories were turned topsy-turvy by this freak accumulation. Who could be certain about any new discovery after this?

Mike Benedum rallied quickly. Transcontinental's long-range purpose was to find more oil, and who was more capable of this than he? As he made plans for a new exploration campaign, he received word that Standard Oil of New Jersey was interested in his Colombian concessions. There was a new sparkle in his eyes as he readied himself for the familiar thrill of pitting wits.

Perhaps it was the cutting edge of failure that sharpened Mike Benedum for the most lucrative bargain he ever made with these skillful buyers. In any case, he was fully aware of the remarkable change that was beginning to take place in the Jersey Standard company.

Death and retirement had lessened the influence of the older executives who had written chapter and verse for John D. Rockefeller's text that control of refining and marketing was the essence of success. Younger men were forging more enlightened policies in Jersey Standard's business and public relations.

Dynamic, personable Walter Clark Teagle became the company's president in 1917. Only thirty-nine, he had ideas to match the bigness of his 6 feet 3 inches and 240 pounds. A Cornell graduate, he was working in his father's Cleveland oil company in 1899 when Standard bought it. Because Teagle possessed remarkable executive talents and a decided flair for harmonious personal relationships, he was rapidly promoted. Within nine years he became a Standard director.

E. J. Sadler, an Annapolis engineering graduate who also joined Standard in 1899 and subsequently managed its principal foreign venture in Rumania, was Teagle's chief lieutenant in changing Jersey Standard's policy. These ambitious young men believed in expanding Jersey Standard's producing activities both at home and abroad. Before they took over, Jersey Standard owned only 1.6 per cent of America's oil production and only 1.9 per cent of the world's. Within three years they had boosted these percentages to 3.2 and 4. Teagle publicized Jersey Standard's intention of going into every producing area in the world.

Appraising this new policy, Benedum recognized that it was motivated in part by the scarcity of oil everywhere. The United States was using up its supplies sixteen times faster than the rest of the world. The great mid-continent fields were rapidly declining. The bright hopes of north central Texas had proved a delusion. Mexico's fabulous production had declined by half. The more Benedum thought about it, the more valuable the Colombian white elephant began to look.

Teagle dispatched a party of geologists to Colombia and obtained a report that the DeMares concession was worth $1,600,000. He knew that Benedum and Trees would snort at such an offer, inasmuch as they had already spent that much in drilling their first three wells. Reluctantly, the geologists raised the valuation to $5 million, pointing out that though the properties were undoubtedly valuable, they were located in impossible country.

Teagle brushed aside his experts' advice. They were conforming, he felt, to the old, cautious pattern that had prevented Jersey Standard from joining Benedum in Colombia four years before on a partnership basis. A million dollars then would have saved whatever the cost was going to be now.

When negotiations started, a Jersey Standard negotiator grumbled, "Mike Benedum sat across the table with a look on his face like a Sunday School superintendent and tried to persuade us that the property was worth five hundred million."

However, to Teagle's surprise, Benedum did not want cash. "Look over your companies and choose one suited to take this project over," he suggested. "Capitalization will naturally have to be increased. We'll take stock in the new company."

The International Petroleum Company, which was developing Jersey Standard properties in Peru, was selected. The negotiators turned to the question of how much of its stock the Colombian properties were worth.

Benedum drove a hard bargain. When some Jersey Standard officials objected to the $33 million worth of International stock the Benedum group received, Walter Teagle's explanation spelled out the difference between Jersey Standard's past and its future.

"There are very few real sure things in the oil game, especially in the producing end," he told his dissenting associates, "and the individual or corporation that does not take some chances never gets very far." It would not be long before Standard's new philosophy would be expanded to include exploration as well as producing.

The Colombian deal would prove more rewarding to both sides than either realized. Instead of producing $500 million worth of oil, as Benedum extravagantly predicted, the DeMares concession went double the sum, producing 500 million barrels of oil. Mike Benedum always referred to his International Petroleum stock as "Old Faithful," for even during the depression of the thirties it doubled its dividend rate.

Six years passed, however, before the first barrel of this abundant new source of oil reached the markets, so great were the difficulties of drilling jungle wells and laying a pipeline. In the meantime, Benedum began to think that, despite his skill as a trader, his luck as an explorer in the United States had run out. Every well Transcontinental drilled was a dry hole.

The dream of empire turned into a nightmarish struggle to salvage the stockholders' investment. As bank after bank refused to loan money to Transcontinental, Benedum began to pour in all his personal resources. Just as he refused to declare bankruptcy when his West Virginia bank failed and eventually paid back all the bank's debts, so he refused to abandon Transcontinental. People had bought its stock because they believed in Mike Benedum's golden touch. He had to justify their confidence. As his resources continued to disappear down dry holes, he did not know that this determination would eventually cost him more than $20 million. He would not have changed his course even if he had known.

When everyone else in the company lost hope, Benedum seemed to tap new reserves deep within himself. "A wildcatter can't quit, no matter how enormous the odds against him," he said simply. "The oil's there waiting, but it won't show itself unless you seek it and seek it strenuously."

But could even such a faith as Mike Benedum's continue to survive such a staggering succession of failures?

3 : A geological brainstorm

$O_{F ALL THOSE}$ whose hopes and ambitions were blasted by the collapse of the Ranger and Desdemona fields, no one was more despondent than Wallace Pratt.

This sensitive, brilliant young geologist knew that the disaster could not have been predicted. He had not lost money personally, but, as in Mike Benedum's case, it was a deep sense of responsibility to others that perturbed him.

Pratt felt that because of his recommendations concerning Ranger the company he was working for had lost millions of dollars and perhaps its opportunity to become one of the nation's bigger companies.

Anyone being told at this moment that Wallace Pratt would become one of the world's greatest geologists would have been incredulous. He was, as one associate put it, "modest to the point of self-effacement." His friends would have scoffed at the idea that he would soon do more than any one man to change the exploration philosophy and policies of Jersey Standard. Pratt himself would have been terrified.

He was painfully timid. As a child in Phillipsburg, Kansas, where his father was a farmer turned lawyer, he was the special prey of bullies. The sixth of ten children, he tried to emulate the aggressiveness of his older brothers, but nothing came easy for him. He had to force himself to do things for which he had no aptitude, physically or psychologically. To earn pocket money he baled hay for a dollar a week. Fighting shyness, he developed a newspaper route and a laundry pickup.

Only in studying did he find a world in which he was at

ease. He loved everything he learned. No work was too hard, no experience too mortifying as long as it meant he was earning money to go to college.

At the University of Kansas, Pratt was so reserved that he impressed none of his teachers. He would probably have remained an obscure student, unaware of his own potentialities, had he not accidentally caught the attention of his geology professor outside of class.

Pratt worked as night clerk at the Eldridge House, Lawrence's leading hotel. Late one night the telephone rang and Erasmus Haworth dictated a telegram—the hotel offering this service after the telegraph office closed. Dr. Haworth said he would pay for it the next time he was in town.

"I'm a student in one of your classes, Dr. Haworth, and I'll be glad to pay for it to save you a trip," Pratt volunteered.

Haworth took down Pratt's name and the number of his classroom seat, but forgot to take care of his obligation.

Several weeks went by. Wallace Pratt, of course, could not bring himself to mention the matter, even though the 25 cents involved was a major sum to him. One day Haworth called on him in class by number and, having lost his card, asked his name. This jogged his memory and he asked Pratt to stay after class.

Once he had noticed the gentle, shy lad, Haworth was attracted by his keen intelligence. With rare understanding, he gave him a job in the geology laboratory, talked with him, encouraged him. Pratt's whole world came alive with purpose. It was as though the roots of a stunted desert plant had finally reached the underground stream for which they had been blindly and instinctively searching. With growing excitement, Wallace Pratt realized that for him the study of the earth was a way of life. He had thought that a man's business of earning his daily bread was labor, simply to be endured; that only by unpleasant struggling could he earn a few free moments to explore and enjoy the beauty of the world, to discover himself through the inspiration of poetry and philosophy. The discovery that a way of life and a way of making a living could be one was a revelation.

James Dwight Dana, America's foremost teacher of earth science, had written a half a century earlier, "The earth's history is the true introduction to human history." As he read this, Wallace Pratt resolved that he, too, would dedicate himself to demonstrating that geology is a philosophic history.

Dr. Haworth was preparing one of the first petroleum geology surveys, *Oil and Gas Resources of Kansas.* Pratt and other students who assisted him learned their geology by participating in creating theories from facts they had to observe for themselves. Like most young geologists of the day, Pratt concentrated on mining and graduated as a mining engineer.

The Bureau of Mines offered him a position in the Philippine Islands. It seemed an exotic end of the trail for the Kansas youngster who knew so little of the world except the stories its rocks told.

Seven years of tropical climate took their physical toll and he resigned, hoping to find work in the States. As he was visiting with Dr. Haworth, a Texas Company geologist dropped into the office, looking for a geologist to go to Mexico. It meant the tropics again, but the assignment was to search for oil. Pratt somehow felt he should not refuse.

From Mexico he was sent to Costa Rica and then to North Texas as division geologist. The Texas Company, developing its rich Electra field, was anxious to discover new ones. To

Pratt's dismay, he received an order to make an immediate location for a wildcat. He searched his office files for reports made by geologists who had preceded him. One survey, indicating an anticline in Eastland County, far south of Electra, intrigued him. After mapping it in detail in the field, he enthusiastically recommended a test.

When the well produced, both Pratt and his company were pleased but not excited—it was not a big producer. They did not realize that they had discovered the Ranger field. The McCleskey farm gusher the following year would capture the title as the discovery well, but subsequent drilling would prove that Pratt's discovery was part of the field.

Wallace Pratt now took, for him, a daring step. He resigned from The Texas Company to buy and sell royalties under prospective oil lands. So good was his geological judgment that some of the royalties were still bringing him an income forty years later. However, the first time he failed to make a sale he lost confidence.

Pratt was a dejected man when he visited Frank Cullinan, a Texas Company vice-president.

"Why aren't you with Humble, Wallace?" Cullinan asked. "They're the best people in Texas. I've told you that dozens of times."

On the same floor with The Texas Company offices in Wichita Falls a door bore the name "Humble Oil & Refining Co." The door was usually open but all Pratt could ever see was a big fat man complacently smoking a cigar.

Someone told him it was R. L. Blaffer, a Humble vice-president. The company, whose main offices were in Houston, had been organized in 1917 by a group of small independent producers who, unable to sell their oil for what they felt it was worth, had pooled their production and talents to buck the major companies who controlled the pipelines and purchased their crude.

The Humble members started their careers at Spindletop. Blaffer left his father's coal business in New Orleans to join the boomers. William S. Farish, a young Mississippi lawyer, was sent by a client to Beaumont and never went back. W. W. Fondren started as a drilling contractor at Spindletop. Ross Sterling, Humble's president and later governor of Texas, set up, with his brother and sister, a series of feed stores in the early boom towns to serve teams hauling equipment. They invested their profits in oil leases and soon forgot the feed business.

At Cullinan's insistence, Wallace Pratt finally generated enough courage to walk through the open door to talk to Blaffer. The fat man chewed his cigar speculatively, then said:

"Well, I guess we do need a geologist. I don't know why, though. We've been doing all right without one. We've got scouts in the field and we've bought some mighty good leases wherever oil's been found. We've got some of the best production in the Ranger field. But I tell you what you do— go down to Houston and talk with Bill Farish. He's the one who thinks we need a geologist."

Embarrassed and dubious, Pratt went to Houston simply because he didn't know how to say no. These might be the best people in Texas, but he didn't feel they were his kind. However, the minute he met Bill Farish he felt as he had when "Daddy" Haworth noticed him. Here was another man who understood. This big, genial, keen-eyed man was a born leader and Pratt was overwhelmed with the desire to follow him, to work by his side. It didn't matter about "the best people." Pratt had found, for him, the best man in Texas.

Farish was equally delighted with Pratt, recognizing the potential greatness of this modest, reserved young scientist. "We are fortunate to get you," he told Pratt. "You know so much about the geology of Ranger and there's more oil to be found there. We'll make you chief geologist. Of course there'll only be you, but why not be chief of your own show?"

Pratt dedicated himself to the study of North Texas geology, convinced that oil-finding problems could be solved only by research.

As the Ranger boom reached its peak, Walter Teagle felt it was time to move into Texas. Jersey Standard's production in Oklahoma and Louisiana was declining rapidly, while the star of Texas was obviously rising. Because of Texas' long antipathy to Standard, concretely expressed in antitrust laws, Teagle deemed it wiser to purchase an interest in an established local company rather than invade enemy territory direct. He first tried to buy the Texas Pacific Coal and Oil Company, the reputed discoverer of the Ranger field. Just before the deal was closed, prolific new wells were brought in and Texas Pacific doubled its price. Teagle refused to buy.

Walter Teagle and Bill Farish had worked together on the Petroleum War Service Committee, and their friendship had grown on hunting trips. Now it occurred to Teagle that the young Humble company might be Jersey Standard's answer.

Farish took Wallace Pratt with him to help negotiate the sale—no one could describe Humble's potentialities or the value of its holdings better than Pratt. Jersey Standard bought a half-interest in Humble for $17 million. Humble was producing 16,500 barrels a day. With its new capital it immediately began to build pipelines and a refinery and to buy more leases, all around Ranger, considered by Jersey Standard to be its most important asset.

Then, in November, 1919, Ranger began to fail. To Wallace Pratt this was a greater cataclysm than any he had studied in the pages of his books or in the earth's ancient rocks. He felt that he had failed his science, Bill Farish, and himself. Humble's golden opportunity was lost. The money was squandered.

"My position with Humble dwindled rapidly and soon all but reached the vanishing point," Pratt said later. "The future looked black and my stock was at a low ebb."

However, Pratt was still convinced that "the winning of oil is a geological enterprise. I believe that oil in the earth is far more abundant and far more widely distributed than is generally realized," he told Bill Farish. "This beautiful conception of finding oil in anticlines is perfectly valid in principle, but I think it is leading us astray in the practical search. A lot of anticlines don't produce oil and a lot of oil is found where there are no anticlines. We just haven't done enough research yet to discover all the different factors that control the accumulation of oil in the earth."

Such talk was fine, of course, but it didn't put any oil in the tanks. Others were finding oil, regardless of how accurate or inaccurate their theories might be. Pratt soon had reason to be even more discouraged.

Albert E. Humphreys, a former Colorado silver miner and Oklahoma wildcatter, drilled a presumed anticline at Mexia, Texas, in November, 1920, and found oil. No one knew how important the field was, but this was new oil territory in south central Texas.

What galled Pratt was that he had told Bill Farish Humphreys didn't stand a chance of finding oil there. After the discovery Pratt spent a week in the field, establishing that the supposed anticline was actually a fault, a fracture in the rock strata caused by earth movements.

Pratt had mapped many faults. Sometimes the movement of the earth caused the strata to slip only a few feet. Sometimes they slipped so violently that on one side of the fracture a rock formation would be found thousands of feet above where it was found on the other side. He had seen dramatic examples exposed in the mountains, looking like a blind giant's attempt to mend a layer cake broken in the middle.

Pratt tried to visualize what must have happened at Mexia millions of years before. When the earth cracked and the strata slipped, on the upthrown side the broken edges of the rock beds were dragged down, sealing them. Later, when oil migrated through the porous sand it reached this barrier and was trapped. On the downthrown side the broken edges were dragged up, and any fluids moving in these beds were free to move on.

The explanation was so simple—and yet wells drilled on *both* sides of the Mexia fault were finding oil.

"I'm utterly distracted," Pratt told his assistant, Dwight Edson. "All the evidence we have is conflicting. It's just wisps of evidence. It doesn't make sense."

They stared gloomily at their maps and notebooks. Maybe if they shook them all up they might fall into a new pattern, like bits of colored glass in a kaleidoscope.

"Dwight, I've just had a brainstorm," Pratt said suddenly. "We've been assuming that this fault is vertical. We've been wearing blinders. Doesn't it explain everything we've found if, instead of being vertical, this fault slants underground? If it dips to the west from where we see a trace on the surface, then wells drilled on what we've been assuming was the downthrown side will naturally find oil. They drill right through the slanting fault and reach the oil sand which is still on the upthrown side of the fault."

Edson agreed enthusiastically. "That means the real producing part of this field is west of where everybody who thinks it is an anticline believes the limits of production will be."

"Yes, and it also means that if we follow the fault line we'll find good production further north. We've got to get Bill Farish here immediately. I don't think anybody else has figured this thing out geologically, but there's a well being drilled west of the fault line on the surface. They think they're on an anticline, but if they find oil, it will prove we're right. Everybody will start leasing to the west, even if they don't guess what we have guessed."

It did not occur to Farish to question Pratt's theory, although the idea that a fault could be the controlling factor in the accumulation of oil was a novel one. The concept of the fault plane dipping west at the remarkably flat angle of 50 degrees was even more unorthodox. Not a geologist in the area could have been found to support Pratt's idea.

Farish, Pratt, and Edson scoured the countryside, leasing every acre available west of the surface fault out to a line Pratt drew. When Farish spent all that Humble could reasonably afford to risk, Pratt pled for more. The well drilling west of the surface fault found oil. Pratt was elated. As more wells produced his theory was proved. A spectacular new field lay 7 miles along the fault, its productive area on the "wrong side," or where it shouldn't have been, according to those who held that only anticlines controlled the accumulation of oil. Humble's fortunes were dramatically restored by Pratt's brilliant deductive reasoning and the hasty leasing campaign that netted the company 10 per cent of the field.

A year later when Pratt presented his new theory before the American Association of Petroleum Geologists, many were still skeptical. Within less than a month Pratt proved it beyond doubt. A small new discovery on a supposed anticline was made 30 miles north of Mexia at Powell. Recognizing that this was another fault accumulation, Humble applied the same rule and managed to acquire a third of the field. Powell's 12,000-barrel-a-day gushers quickly reached a daily production of 354,893 barrels—more than Cushing produced at the height of its glory.

Although both Powell and Mexia quickly declined, they produced more than 172 million barrels in five years. Humble was now firmly established as the largest producing company in Texas.

Curiously, Powell lay only 8 miles from Corsicana, where Texas' first oil was found in 1894. The area's oil reserves had long been considered depleted.

Pratt's success intensified his belief that a geologist should never take anything for granted. He was fond of telling how indignant he and other Kansas geologists were in 1915 when the driller of a wildcat reported finding granite at 1,100 feet right in the middle of the state. In their minds it was an impossibility to find this basement rock so close to the surface in Kansas.

A group of geologists visited the well and when the driller bailed out fragments of beautiful pink granite, they claimed he had salted the well with it. For what purpose, they did not explain. As far as the driller was concerned, finding granite meant the end of his chances to find oil at that spot. It took several years and many more such granite discoveries to make the geologists realize that there was a great range of buried mountains crossing the width of Kansas just under the surface and that it had been a prime factor in forming oil traps along its flanks.

"As geologists we like to emphasize the contribution we make to the oil industry," Wallace Pratt said. "Our effort to make geology serve industry has really benefited geology more than industry. The bread we have cast on the waters has come back to us many-fold. We have learned far more than we originally had to impart."

In the 1920's geologists still knew so little that the more positively one might state that oil could not be found in a certain place or under a certain condition, the sooner he would probably be proved wrong. The profession gleefully recalled the example of Dr. W. Van Holst Pelekaan, Shell's chief geologist in California.

Late in 1919 Pelekaan went to Peru and Ecuador to make a survey. While he was gone, D. H. Thornburg a Shell geologist who had grown up in Long Beach and remembered the tilted beds he had seen as a boy, mapped an anticline on Signal Hill, which rose sharply from the beach and had been used as the site of a beacon for ships. He recommended drilling a well.

Dr. Pelekaan was outraged when he learned that such bad judgment was prevailing. Union Oil had already drilled a dry hole on Signal Hill. As he hastened back to California, another Shell geologist asked him to stop in Salt Lake City to make an inspection trip through western Colorado and eastern Utah.

"I haven't the time," Pelekaan said. "A damned fool has

got us drilling a well in Long Beach and I've got to get out there to stop it."

Dr. Pelekaan arrived just in time to receive the news that on June 25, 1921, the well on Signal Hill started flowing 600 barrels a day, opening America's fourteenth largest oil field, containing 926 million barrels.

Wallace Pratt, recognizing that the infant science of geology was still tangled up in its swaddling clothes, persuaded Humble to start research in micropaleontology, for he felt that a key to petroleum geology lay in the accurate identification of the age and nature of rock formations by means of the tiny fossils they contain. The Humble research laboratory began studying rock cores from many wells, correlating formations over wide areas to obtain a picture of underground structures.

Pratt, given a free hand to carry out his exploration ideas, felt that his future was no longer black.

4 : Indian prophecies

THE CROWD GATHERED early under the big elm tree on the Osage Tribal Reserve in the town of Pawhuska, Oklahoma. The brisk November air fairly crackled with excitement. Saturday was always town day but this morning it seemed that all the ranchers and farmers in the Osage hills had converged on this one spot. The whites, accustomed as they were to their Indian neighbors, stared with curiosity at certain old full bloods and their squaws who were seldom seen in public. Squatting on the lawn, they were a quiet, dignified pool around which the boisterous crowd eddied and rippled.

The full bloods were a colorful group. The old men were richly and warmly clad in rainbow-hued blankets, buckskin leggings, and beaded moccasins. Their iron-gray hair hung in long braids over their shoulders. Some wore the traditional beaver-skin bandeaux, reminiscent of Cossack caps, while others wore black Stetsons decorated with eagle feathers. Their earrings were of gold. An occasional elaborate bear-claw necklace could be glimpsed under a blanket. Some of them clutched an eagle-feather fan despite the day's coolness. Their squaws were more decorously dressed and blanketed.

Only those who knew them well could detect that the full bloods were enjoying the excitement as much as the younger mixed bloods who were chattering away, now in Osage and now in English, among themselves and with their white friends.

Although the crowd had gathered for another reason, news had just arrived that this very morning, November 9, 1918, Kaiser Wilhelm had abdicated. The German armistice commission was meeting with Marshal Foch in his railway carriage. Many sons of the tribe would soon be coming back from overseas.

Talk of the armistice suddenly ceased as a large group of well-dressed businessmen approached. The special train from Tulsa had arrived. This was the moment everyone had been waiting for. The auction of the Osage oil leases was about to begin.

As the oilmen shoved their way through the crowd, trying to get as close to the auctioneer's stand as possible, the full bloods watched them closely. The expressions on the oilmen's faces were as inscrutable as those of the Indians. The oilmen scanned the crowd to see what other competition had arrived by private automobile or from other cities.

All the big concerns were represented—Carter Oil Company, Jersey Standard's subsidiary; Gypsy, Gulf's subsidiary; Roxana, the Royal Dutch-Shell company. Harry Sinclair's man was there, and there were dozens of independent operators, each sizing up the other's bank account according to the boom in which he had made it. There was W. G. Skelly—Healdton and El Dorado behind him. There was Waite Phillips, strikes in several rich little fields to his credit. His brother Frank was present on behalf of the newly organized Phillips Petroleum Company, capitalized at $3 million. There were those who had prospered at Glenn Pool and at Cushing —and there were those with only a few hundred dollars who had not yet found oil but were hoping that by some miracle they could snatch a small bone from under the noses of the well-fed members of the pack.

The game would be played for high stakes today. Three days before, the American Pipeline Company, owner of a tribal gas lease, had brought in a well that was expected to find gas. Instead, oil had flowed over the derrick at the rate of 5,000 barrels a day, quickly increasing to 9,400. It was Oklahoma's biggest gusher since Cushing, but the company that drilled it had no right to the oil. The lease was among those to be auctioned this morning. How much, men wondered, would a gusher be worth on the block?

There was a gleam of pleasure in the eyes of some of the old full bloods. Doubtless they were wishing that their Chief Wah Ti An Kah could have lived long enough to see how wrong his prophecy was.

In 1872, when the tribe was forced to sell its lands in

Kansas and buy a reservation elsewhere, it was Wah Ti An Kah who had urged them to come to these hills instead of going to the western part of Kansas where there were still buffalo herds.

"White man will not come to this land," he declared. "White man does not like country where there are hills, and he will not come. White man cannot put iron thing in ground here. This country is not good for things which white man puts in ground. This will be good place for my people."

So the Osages bought 1,500,000 acres in the blackjack-covered hills. The land they bought from the Cherokees had once been theirs. By right of ability to defend it from other Indian tribes the Osages had owned all the territory that later became Arkansas and Missouri. The Osages were fierce, proud, courageous, aggressive. They were also handsome. Their men averaged over 6 feet in height.

Wah Ti An Kah had been right in thinking that this hilly country, their last refuge, would never attract the white man's plow, but he knew nothing of that other "iron thing," the drilling bit. In 1896, a dreamer from Rhode Island, Edwin B. Foster, learned of springs that the Indians skimmed for medicinal oil. With government approval, he obtained a ten-year lease on the entire Osage reservation. After drilling two dry holes, he found some oil in his third. But he was too soon. There was no way to market the oil and his capital was gone. In 1902 he assigned his leases to the Indian Territory Illuminating Oil Company, which Cities Service bought a few years later.

The original oil lease of the entire reservation was renewed for another ten years. When it expired in 1906, a thousand wells had been drilled, discovering a number of small oil and gas fields. The Indian Territory Illuminating

Company was permitted another ten-year renewal, but this time the lease was restricted to 680,000 acres in the eastern portion of the reservation.

Although the Osages were forced to accept allottment of their lands in 1906, to be ready for statehood, they were able to obtain better conditions than the other tribes because they had purchased their lands. Each of the 2,229 members of the tribe received 657 acres of land, but the oil, gas, and other mineral rights, at their chiefs' insistence, were reserved for the tribe as a whole.

When the Indian Territory Illuminating Oil Company's lease expired in 1916, it was permitted to keep any 160-acre tracts whose wells were producing less than 25 barrels of oil a day. The balance, 15,000 acres producing almost 5,000 barrels a day, was sold at public auction for $3 million.

Exploration and development had been steady if not spectacular. The whole tribe prospered because the lease-sale bonuses and the proceeds of royalty payments (one-sixth of the oil produced), were divided equally among the legally enrolled members. An Osage might squander or be swindled out of his quarterly payment after it was in his pocket, but the source of his income was safe in tribal ownership.

As the tribe gathered under the elm tree to watch the November, 1918, lease sale, each member had received $3,600 during the previous year—the 3,755 wells on their lands had produced almost 11 million barrels of oil. One family, with nine enrolled children, had received $39,600.

As Wah Ti An Kah had said, for the wrong reason, "This will be good place for my people." Now, with a 9,000-barrel gusher on their land, it looked as though it would be even better.

A rugged, big-framed man shouldered his way through

the crowd and took his place on the auctioneer's stand. He posed dramatically with upraised gavel. Talk died away. In a resounding voice, Col. E. E. Walters opened the sale.

No matinee idol ever had such enthralled audiences as Colonel Walters. He had been a small-town auctioneer of livestock, town lots, and farms until the opportunity to auction the Osage leases revealed his great talent. Building suspense and warming up the bidders, he started off with a number of tracts far from the tantalizing jackpot lease. Then, in a bored manner, he asked, "Now where will we start Tract 6?—in case anybody would be interested in bidding on it."

This was the gusher location. There was a roar of laughter and only Walters caught an opening bid from a man in the front row.

"I'm offered a hundred thousand," he boomed. "Who'll make it two? Two. Who'll make it three?"

The Colonel moved so fast that not even the oilmen could tell whose nod his sharp eyes caught. "Six, I'm offered six. Who'll make it seven? Seven? Seven? Sold to Gypsy Oil Company for six hundred thousand."

The full bloods exchanged satisfied nods. The bonus on the single 160-acre gusher tract had brought them fifteen times more money than all the 25,000 acres leased at their first auction.

The gusher sale generated a gambling fever, and the next tract sold for $17,000. A tract adjoining a dry hole went for $40,000. When Colonel Walters had worked his way through 94 leases he had enriched the Osage tribe by $2,712,600 in bonuses. The perspiration was rolling over his celluloid collar, but he had not yet finished earning his $10 fee.

Until today only the eastern part of the reservation had

been opened for leasing. Now 175 tracts in the unexplored western half were to be offered.

A few dry holes had been drilled by the Indian Territory Illuminating Oil Company in this part of the reservation before it was closed to leasing. The opinion among oilmen was that only the eastern half was good hunting ground. The crowd began to thin when Colonel Walter's persuasiveness could bring only $500 or $600 for the first tracts. Those who remained heard Colonel Walters saying again and again, "Sold to Marland Oil Company."

A stubby, imperious man who had not bid on the eastern leases had quietly moved to the front and was bidding with little or no competition. The representatives of the big companies watched him speculatively. E. W. Marland had discovered the big Ponca City field west of the Osage reservation, and he had assembled a geological staff that outnumbered those of many of the big companies. Several of his geologists were by his side.

As the location of the tracts shifted, Marland consulted a map and made no more bids. Nineteen tracts were sold before he bid again. This time Carter and Gypsy men began to bid against him. The bid ran up to $8,500. When Marland imperturbably bid $9,000, they shrugged their shoulders. But a battle had started. Marland got no more leases for $500. He was forced to bid as high as $26,000 for the leases he was obviously determined to have. There was a definite pattern to his bidding. You could almost visualize structures on his map from the tracts he bid on.

When the sale ended E. W. Marland and Carter Oil had bought the greatest number of west side leases, and an additional $646,000 was in the Osage treasury.

E. W. Marland and his staff were in a mood to celebrate as they returned to Ponca City. They started a poker game but it was difficult to keep their attention on the cards with the excitement of the real game so fresh in their minds.

"What a break for us, E. W., that everybody else was excited about the gusher," Spot Geyer, Marland's chief geologist, said.

"Yes, but I still had the advantage, even when they started to give me a little competition," Marland said complacently. "Those major company boys are told before they get to Pawhuska just how high they can bid on any lease. They haven't got a chance to win if it's something I really want."

Spot Geyer unrolled the well-thumbed map. "When the next sale comes up in March, I'll bet it's going to cost you plenty to get the rest of the leases we want. They know we're after something and they'll be authorized to bid high."

Marland waved the map aside with an expansive gesture. "If it's as big as we think it is, it'll be worth all I'll have to pay for it. They're not real gamblers. They just act like it when they think they've got a sure thing."

Marland watched his enthusiastic young assistants with genuine affection. He was proud of their brains and ability, but he knew it took more than knowledge to be a big winner. It took the gambler's special courage to keep on playing for high stakes, even when the game was going wrong.

When Ernest Whitworth Marland came to Oklahoma from Pittsburgh in 1908, he had already made and lost a million dollars. No one could have told by looking at this natty little man in knickerbockers, Norfolk jacket, and spats that he was living on borrowed money. His every word and mannerism gave the impression that affluence was his birthright. He had inherited from his father, Alfred Marland, the social snobbery of a middle-class Englishman, adding to it the American reverence for wealth. He also inherited the ambitious restlessness that had caused his father to quit England and the teaching profession to enlist in the Confederate army.

After the Civil War, Alfred Marland prospered in Pittsburgh by inventing and manufacturing an iron band to bale cotton. He raised his son in the English tradition, even sending him to secondary school in a curious English colony established in the Tennessee mountains by Thomas Hughes, author of *Tom Brown's School Days* and *Tom Brown at Oxford.*

Alfred Marland wanted his son to be a statesman, but as a law student at the University of Michigan, E. W. distinguished himself chiefly as a poker player.

When his father's modest fortune did not survive the depression of the 1890's, E. W. decided that aristocratic ideas and ideals were of little value without money. He had no idea what the formula for becoming a tycoon was, but surely a gentleman could become one easily.

While he imagined himself moving in the same financial circles as Andrew Mellon and Andrew Carnegie, the young Pittsburgh lawyer earned his living checking coal leases for the team of Guffey and Galey. He quickly realized there was money to be made in buying and selling options on prospective coal lands. It was not long before his successful promotions and the lavishness with which he spent whatever money he made made Pittsburgh businessmen aware that E. W. Marland was on his way up.

Fascinated with the thought of how quickly he could make a fortune if he found a rich new vein of coal in the depleted area close to the steel mills, Marland looked into geology.

He was delighted rather than disappointed to learn how inexact the science was. It seemed quite possible that a rich deposit could have been overlooked.

The young lawyer was no scientist, nor would he ever become one. He was too impatient of detail and lacked the true scientist's motivations. However, as he taught himself the rudiments of geology, he developed an understanding of science's practical uses. It was to reward him richly.

After examining outcrops and peering into ancient diggings Marland was convinced he had found an extension of a rich coal vein from Pennsylvania into West Virginia. Jubilantly, he leased 5,000 acres and hired a contractor to drill a core hole. They found several thin strata of coal. Marland insisted on going deeper for the bonanza which had to be there. It was there, all right, but it wasn't coal. At 747 feet he struck oil, 100 barrels a day.

Although he began with coal leases, he managed to acquire a few oil leases in the heart of what proved to be the Congo field. The field was small, but oil and gas so close to a big market brought him his quick fortune.

It vanished as quickly as it arrived. Within a year the field was depleted and the national panic of 1907 brought economic chaos.

Calling on Marland, a friend asked how he was doing. He pointed to a bill hook. "There, on that hook, are one hundred and fifty thousand dollars' worth of unpaid bills." Marland also had the responsibility of a wife, an attractive Philadelphia girl, Mary Virginia Collins, whom he married during his first flush days as a coal-land promoter.

Help arrived from an unexpected direction and in a more romantic form than even Marland's extravagant imagination could have conceived. A nephew, Franklin Kenney, an Army lieutenant stationed at Fort Sill, Oklahoma, had become a good friend of the Miller brothers, Joe, George, and Zack, sons of an old-time cowman who leased a vast ranch from the Ponca Indians. The brothers were organizing their 101 Ranch Show, which soon would become a world-famous circus. Too busy to think about oil in their part of the state— where none had yet been found—they were agreeable to Lieutenant Kenny's suggestion that his "rich" oilman-uncle look into the possibility.

Marland didn't even have money for the train fare, but the glamor of oil, Indians, and Oklahoma, where the giant Glenn Pool boom was in full swing, was enough to promote some capital from friends of his more prosperous days. The fresh, informal community of Ponca City burgeoning on the rolling prairie was captivating. Marland viewed the strange conglomeration of cowboys, Indians, and circus performers as characters in a wonderful story. He would now write himself the hero's role. Here he could build an empire in which he would be a benevolent ruler, free from competition with the Eastern tycoons whom he had envied so long.

With his amateur geologist's eyes, Marland saw anticlines wherever he looked. He located his first well near the Miller brothers big white ranch house. The drilling equipment broke down constantly. The hole inched down. Marland put his last thousand dollars in the bank, instructed the driller to draw on his account to meet the payroll, and returned to Pittsburgh to raise more capital.

The second check the driller wrote was returned marked "Insufficient funds"—Marland had been the loser in a Pittsburgh poker game. However, his spellbinding tales of the wealth lying under the prairies coaxed more money out of his Eastern friends.

The first well was abandoned. Marland drilled seven more, finding a little gas but no oil. As optimistic as ever, he charmed Mrs. Annie Rhodes, owner of Ponca City's Arcade Hotel, into accepting his gold watch as security for his board bill. This would have humiliated E. W. Marland, the aspiring Pittsburgh tycoon; it bothered the Oklahoma oilman not at all. When a man is exploring for oil, the only reality is the next wildcat, the one that will come in. He lives so completely in his undiscovered wealth that the struggle to pay his bills is what seems like a dream.

Marland found a new source of capital and prepared to drill another wildcat. As he rode across the prairie searching for a likely location, a long hill rising in eloquent loneliness attracted his attention. Approaching it, he saw that it was topped by four poles from which swung above the reach of coyotes and wolves, the wrapped bodies of Ponca Indians. The hill was the Ponca burial ground where the Great Mysteries would be sure to see a departing spirit placed so conspicuously to attract His attention.

What attracted Marland's attention was the anticlinal shape of the hill. He had an immediate conviction that this was where he must drill. White Eagle, chief of the Ponca, reluctantly gave his consent to an oil lease.

"You are making bad medicine for the Ponca and bad medicine for yourself," he warned Marland.

It would not be many years before White Eagle's prophecy would come true, but Marland could foresee only good medicine in the exciting prospect of finding oil. This time he won the jackpot. There was a great reservoir beneath the burial hill.

Now that his losing streak was broken, Marland raised the stakes; as he developed the Ponca City field, he sent rigs to new wildcat areas in the Ponca lands, discovering two more pools.

One night in the Arcade Hotel, he met Dr. Irving Perrine, head of the University of Oklahoma's geology department. The self-taught geologist began to check his theories with those of the professional. Dr. Perrine enthralled him with new insights into the knowledge petroleum geologists were beginning to accumulate. Marland offered to give vacation jobs to all his students who wanted them.

"I will pay them well," Marland said. "I want them to have the benefits of working here in the field—to learn from the best teacher, the earth."

Marland glowed with pleasure as he thought of the young geologists who had assisted him at the Osage lease auction. Some of them were Perrine's students. He had given them their start. Now they had convinced him, even before he drilled, that the western part of the Osage must contain tremendous quantities of oil.

Marland pampered everyone who worked for him. He liked to provide the luxuries he felt were essential for a man's freedom and growth. His palatial mansion was more like a club for his employees than a residence. Formal gardens, swimming pool, and golf course had all been created on land that had been the town refuse dump a few years before.

Now the Marland estate was one of the garden spots of the Southwest, planted with rare trees and shrubs from all over the United States. All this was just a beginning—nothing compared to what Marland planned to create with the new wealth he would bring up from the ground. And Marland felt that it should belong not just to him, but to everyone who had participated in finding it. He was the person-

ification of Lao-tse's definition of generosity: "He who ob-
tains has little, he who scatters has much."

Competition was keener at the next Osage lease auction;
Marland had to pay $400,000 for the leases he wanted. Mean-
time his rigs were busy drilling his east-side leases. As new
discoveries were made on these, he postponed his exploration plans for the west side. Another irresistible auction was
coming up in October at which 160-acre tracts adjoining
the new east-side production would go on the block.

What poker game could offer the thrill of bidding against
Jersey Standard, Shell, and Gulf, raising them $100,000 at a
time with a nod of the head? The winner might very well be
the loser, for only the drill would reveal whether the lease
contained enough oil to justify the price. Dry holes or small
wells were drilled with disappointing frequency adjacent to
rich leases.

The October auction was spectacular. Carter, Jersey
Standard's subsidiary, came prepared to take the cream of
the leases, but Marland recklessly topped every bid. He set
a new record for a bonus on a single 160-acre tract, paying
$620,000 to top Carter's last bid of $610,000. He captured
another lease for $571,000 and another for $545,000. Of the
$6,161,500 paid in bonuses that day to the Osages, $2,477,300
came out of Marland's pocket. Marland would not have cared
if the leases were nonproductive, so great was the satisfaction
of publicly and dramatically beating the Rockefellers and
the Mellons.

He had another reason to be elated. Deciding to drill
deeper in the Ponca City field, his geologists had discovered
a rich new producing sand with wells gushing 3,000 and
4,000 barrels a day where before they yielded only a few
hundred.

Not until December did Marland finally start a wildcat on his west-side leases. There was only casual interest in the announcement, thanks mainly to a statement made a few months earlier by Dr. David White, chief of the United States Geological Survey. Noting that the nation was looking to the Osage territory for new production, Dr. White expressed doubts that any great field would be found there. He concluded gloomily that within a few years America would be compelled to buy its petroleum and gasoline from the English, who seemed to have the only good potential oil lands under concession in other countries.

Marland read the statement with amazement. How could a scientist say such a thing? He should be saying just the opposite. Each great new pool—Glenn Pool, Cushing—was a proof of abundance in the earth, not scarcity.

"Dr. White has a big surprise in store for him," Marland remarked to Spot Geyer as the west-side wildcat started down.

And one was in store for E. W. Marland. His geologists had picked an anticline to drill and it was the producer they expected. But as they and other companies drilled more wells to determine the size and extent of the discovery, they learned that the surface anticline had nothing to do with the presence of oil. The oil was in a blanket sheet of sand, 50 to 80 feet thick, lying about 3,000 feet below the earth's surface, gently dipping from northeast to southwest. An impervious barrier of shale on the northeast prevented the oil from migrating further.

The geologists had discovered another giant oil field but for the wrong scientific reason. Oklahoma had surprised the nation with the Burbank field, a giant bigger than Glenn Pool and almost as big as Cushing.

Marland and his geologists, judging simply from the mistaken surface clues, had no way of knowing what a great area the field would cover. As discovery after discovery was made, the Osage sales in the Burbank field reached fabulous heights, companies bidding as high as $1,990,000 for one 160-acre tract. The bonus on this one lease was almost twice as much as the Osage tribe had originally paid the Cherokee nation for the entire reservation of 1,500,000 acres.

Dr. White's pessimistic view was at least right in part: the nation could indeed look to the Osage for a great share of its oil. By 1958 the reservation had produced 709,500,000 barrels, with 740,000,000 barrels remaining to be pumped to the surface, and the entire reservation was not yet thoroughly explored. The canny Osages enriched themselves by $400 million. Unleased tracts continued to lure oilmen to Pawhuska for the familiar auctions for many years.

Marland's young geologists had hardly finished congratulating themselves on the Burbank discovery, and condemning their lack of knowledge which prevented them from owning the entire field, when they recommended the kind of prospect which particularly appealed to Marland.

The Tonkawa anticline, 34 miles west of Burbank and only 12 miles west of Ponca City, was well known, and several dry holes had been drilled in the area. Their wits sharpened by Burbank's geological surprises, Marland's men decided that other geologists had written off Tonkawa's possibilities too hastily. In June, 1921, just a year and a month from the date of his Burbank discovery well, Marland discovered another giant oil field. His Tonkawa wildcat came in at 1,000 barrels a day.

At last Marland could be lavish on a princely scale. With his oil holdings valued at more than $100 million, he began to spend money as though he owned the oil resources of the entire planet. He gave his associates great blocks of Marland Oil Company stock, paid unheard of salaries and bonuses to his executives. He bought life, accident, and health insurance policies for all his employees, gave them free medical and dental services, built them elaborate recreational facilities. He bought land and built them two hundred houses on 1-acre tracts.

He gave money to everyone he could think of—Boy Scouts, Girl Scouts, churches of all denominations, orphans' homes, universities. He built a hospital and an athletic field for Ponca City. He established and stocked a 300-acre wildlife preserve, built lakes and stocked them with fish. He converted the preserve's rock quarry into a public swimming pool. He gave Ponca City land and $100,000 for a building to be used by the Masonic Lodge, the American Legion, and a young men's club patterned on the YMCA.

There was no distinction between the munificence he showered upon others and the pleasures he created for himself. Giving to others was as great a personal enjoyment as his yacht, his private railroad car, and his beautiful estate, which he now enlarged to cover 400 acres.

He brought polo to Ponca City, building several polo fields and offering the use of his specially bred ponies to anyone who wanted to play. The Marland Oil Polo Team played with distinction against polo teams from nearby Army posts.

Marland also imported English hunters, hounds, and foxes. Oklahomans dutifully donned red coats and rode to the hounds, but basically were as bewildered as one memorable imported fox. The animal was driven to the scene of the hunt and turned loose. When the hunt began, the pack circled widely and returned to the car that had brought the fox. The English master of hounds was in despair over the stupidity of his pack until, riding up to the car, he saw the fox curled up on the front seat, refusing to have anything to do with a country that belonged to coyotes.

Marland was working as hard and as expansively as he was playing. He sent his geologists and drilling rigs to Louisiana, Colorado, and Mexico. He enlarged his refinery and built a chain of service stations. His prodigality in business and pleasure made it necessary for him to sell some of his leases. He formed a partnership with Royal Dutch-Shell on three of his Burbank leases and then on his Tonkawa holdings. It was an agreeable arrangement; Marland liked the Shell group and they liked him. They were gentlemen—he felt he was among his own kind in doing business with them. Furthermore, he thought that Shell was far in advance of any American company in exploration and production methods.

When Marland's wells couldn't produce fast enough to give him the cash he needed, he proposed a merger with Shell, involving all their American properties. Both groups had the same objective: a completely integrated company with production, pipelines, and refineries on a big enough scale to take the markets away from the Standard companies. Shell thought the proposal admirable, and was willing for Marland to be president of the new American company. However, when it came to working out details, Marland put too high a value on his company. With regrets on both sides, negotiations were broken off.

Marland's regrets would soon be profound. At present he was intoxicated with success. He felt he was just beginning. He wanted to explore for oil in California, New Mexico, and

Texas. Of course he would need money, but that would be no problem.

As it happened, in 1923 money sought him. J. P. Morgan, of Morgan and Company, suggested that he become Marland's banker. It did not occur to Marland to wonder why. He was flattered that at last he was being honored as an equal by the great financiers. Long ago he had dreamed of this day. He was delighted to sell Morgan $12 million worth of Marland Oil Company stock.

E. W. Marland did not know it, but the bad medicine Chief White Eagle had predicted was beginning to brew and bubble.

5 : *"Let the horns go with the hide!"*

IN THE EXUBERANT CROWD gathered at the Kentucky Derby in May, 1923, no one was happier than Harry Sinclair. His horse, Zev, had won. As he accepted congratulations from excited friends and strangers, it seemed to him that the thrill of this moment symbolized the ultimate triumph toward which he was moving so steadily. What in life could compare with proving that you were a winner in the stiffest sort of competition? Today he could feel almost complacent about his accomplishments. He was not yet bigger in oil than Rockefeller but the Standard companies now recognized him as their most powerful competitor. One of them had even joined him.

Finding oil had never given Sinclair the pleasure he had just experienced in making a deal with Standard Oil of Indiana, the supreme refining and marketing organization in the Middle West. With headquarters in Chicago, it depended principally on another old Standard company, Prairie Oil and . Gas, to supply and transport the increasingly great quantities of crude oil it needed.

When Sinclair built a line from the Cushing field to Chicago, he transported crude for Standard of Indiana. When he raised charges, Standard of Indiana announced it would build its own pipeline. Sinclair shrewdly offered them a half-interest, which they bought for $16,390,000 cash. As a pipeline is no good without oil, the Sinclair Crude Oil Purchasing Company was capitalized for $20 million to buy oil for both Sinclair and Standard of Indiana, the new company's joint owners.

Harry Sinclair was now playing at the same poker table as the big boys. Furthermore, he had adopted their philosophy: the safest, most lucrative way to play the game was to buy other people's oil. Sinclair had made the transition from oil finder to financier. His home and headquarters were in New York now, the nerve center of the big deals, the big money. If it wasn't a big deal, Harry Sinclair wasn't interested. He was feeling particularly pleased with the big one he had just closed in Wyoming. He was sure he was sitting on top of 130 million barrels of oil snatched out from under Standard of Indiana's nose.

By purchases and mergers Standard of Indiana had control of the Salt Creek field in the dreary, inaccessible flats of Wyoming. Although oil was first discovered here in 1909, it was not until 1921 that development by numerous small companies reached such proportions that Salt Creek was recognized as another giant oil field, almost as big as Cushing.

The south end of the Salt Creek anticline, crossed by Teapot Creek, is known as Teapot Dome. All the area was public land until 1915, when the government set Teapot Dome aside as Naval Petroleum Reserve No. 3. Salt Creek field development soon reached the Naval Reserve boundary line and the Navy began to fear that its oil was being drained by private operators.

There was an even more serious drainage problem in its two other Naval reserves in California. There, some private claims were producing inside the reserved land.

In 1920 a worldwide shortage scare made the Navy highly nervous about its fuel supplies. It was so short that the nation witnessed the amazing procedure of Secretary of Navy Josephus Daniels as he ordered troops landed in Martinez, California, to seize by force 2,000 barrels of privately owned fuel oil.

The Navy also had a case of jitters about international storm warnings flying in the Pacific; it wanted to build up Pearl Harbor as an important base to protect the Pacific Coast. Unless large amounts of fuel oil could be stored at Pearl Harbor, the project would be useless.

All these factors made it logical for the Navy to lease its reserves for development with the option to take its royalty payments in fuel oil and products. The Navy had neither experience nor personnel to handle such contracts, but the Department of the Interior had both in its Bureau of Mines and United States Geological Survey. At Navy Secretary Edwin Denby's request, President Harding issued an executive order committing the administration of Naval oil reserve lands to the Secretary of the Interior but retaining control of policy in the Secretary of the Navy.

The government, still hostile to the Standard companies, wanted strong independents to develop the reserves. Admiral J. S. Robison, as Secretary Denby's representative, and

Secretary of Interior Albert B. Fall, invited Harry Sinclair to make a proposal concerning Teapot Dome and Edward Doheny to make one for the California reserves.

The Navy drove a hard bargain. For the privilege of developing Teapot Dome, it demanded a royalty of up to half of all the oil produced, a higher price than Standard of Indiana paid for oil in Wyoming and a $22 million pipeline to be built from Wyoming, connecting the field with the mid-continent pipeline systems.

Sinclair's associates argued that the contract would be ruinous. He decided to withdraw and returned all the papers to Secretary Fall. Fall and Admiral Robison refused to take his refusal seriously.

"Well, if we're ruined, we're ruined," Sinclair said. "Let the horns go with the hide!" He signed the lease.

The magnitude of the deal appealed to him, and he reminded his associates that the Bureau of Mines estimated Teapot Dome contained 130 million barrels of oil. If this proved to be so, there could be no doubt about making a profit.

In return for the California lease, Doheny agreed to an equally high royalty and the option to pay it in fuel oil. Further, he agreed to build the Navy 1,500,000 barrels of storage capacity in Pearl Harbor. He, too, was flattered by the magnitude of the enterprise and the special glitter of such an important government contract. Nor was he modest about the profit he expected:

"If we don't make a hundred million out of this, we'll be in hard luck," Doheny declared. Unlike Teapot Dome, the California reserves were already producing, and Doheny knew from experience how rich a California layer-cake oil field could be.

The contracts were signed in 1922. Sinclair immediately began building his pipeline and drilling wells. His optimism was reflected in the name he selected for the company to operate the contract—the Mammoth Oil Company.

As Sinclair and his party celebrated Zev's victory at Churchill Downs in May, 1923, he thought with pleasure of the 4,000-barrel-a-day production that Mammoth had developed at Teapot Dome. He was also delighted about the trip he was going to make to Russia. The Russian government had expropriated all foreign oil properties, but an enterprising American, George Mason Day, had obtained a contract for a number of concessions in the Baku area. Sinclair was already operating in Mexico—perhaps Russia would be his door into the Eastern Hemisphere. He was taking with him Albert Fall, who had resigned two months before as Secretary of the Interior and returned to his New Mexico law practice. An ex-Cabinet member would be an impressive addition to Harry Sinclair's entourage.

Sinclair was now forty-seven. When he made his favorite comment, "All I have is time and youth ahead of me," he meant it now more than ever. He felt his powers driving him forward to the achievement of his goal.

If there were warning signals of the coming catastrophe that would almost destroy him, he did not heed them. The year 1923 was his last golden one. As 1924 began he found himself a central figure in the greatest political scandal in the nation's history.

In November, 1923, the Senate Committee on Public Lands opened an investigation of the Naval Reserve leases to Sinclair and Doheny. A jealous neighbor reported that Albert Fall had made lavish additions and improvements to his New Mexico ranch. As the committee pried into the

source of his sudden wealth, Democrats delightedly proclaimed that they had opened the lid of a political Pandora's
box which would overthrow the Republican administration
in the next election.

Fall informed the committee that he had obtained a
$100,000 loan from Washington newspaper owner, E. B.
McLean. McLean denied this under oath. Then Doheny
appeared on the witness stand and said he had loaned the
money to Fall just as a favor to a friend.

"It is no more of a sacrifice to me to let a friend have a
hundred thousand dollars than it is for many men to part
with ten or fifteen dollars," Doheny commented. Doheny's
son had taken the money in cash to Fall, carrying it in a
little black bag.

The national gasp over this revelation scarcely subsided
before Col. J. W. Zevely, Sinclair's attorney (after whom his
Kentucky Derby winner was named), testified that Sinclair
loaned Fall $25,000 in the summer of 1923, and it had not
been repaid.

The Democrats adopted little aluminum teapots as campaign badges. The Senate chamber echoed with political
diatribes. "I ask how much more infamous it is to have sold
every gallon of reserve oil than it was for Benedict Arnold
to try to sell only a rock fortress on the Hudson river?"
shouted a Democratic senator from Arkansas.

Doheny's "black bag" became a national joke. In Egypt,
King Tutankhamen's burial chamber had just been opened
after a year's speculation as to what treasures it would contain. A nationally syndicated cartoon showed an explorer
lifting out of the tomb a black bag labeled "E. L. Doheny's
Satchel." The caption read, "What we really expected to
find."

The committee hearings, and the civil and criminal trials which resulted, dragged out for six years. The facts were so obscured by political and editorial oratory, and legal and technical points, and so many side issues were brought up, that it is doubtful the public ever clearly understood what had happened.

Key witnesses fled the country. The principals were bafflingly uncooperative. Bit by bit a broken mosaic, approximating a picture of the happenings, was pieced together. By this time a bewildered public, impatient for tomorrow's headlines, had lost interest.

After six years of investigation the situation boiled down to this:

1. The scandals did not affect the Republicans' popularity. Calvin Coolidge was elected President in 1924 by a huge majority.

2. Civil suits, charging conspiracy to defraud the government, against both Sinclair's Mammoth Oil Company and Doheny's Pan American Oil Company, were dismissed by district courts after lengthy hearings. However, the Supreme Court reversed these decisions with the perplexing comment that "the contracts and leases and all that was done under them were so interwoven that they constitute a single transaction not authorized by law and consummated by conspiracy, corruption and fraud." The three Naval Reserve leases were canceled.

3. Sinclair was cited for contempt of the Senate and sentenced to three months in jail and a $500 fine because he refused, on advice of counsel, to answer one question in a Senate committee hearing, on the grounds that a criminal charge had been filed against him by the government and that an answer to the question had no bearing on the committee's purpose of studying the need for new legislation, but might disclose information affecting his defense in the forthcoming trial.

4. In the criminal action, on the charge of conspiracy to defraud the government, Sinclair was tried twice. The first trial resulted in a mistrial after Sinclair was accused of employing a detective to shadow jury members. In the second trial the jury acquitted him unanimously on the first ballot, being out of court only forty minutes.

5. Sinclair was charged with contempt of court for shadowing the jury, a common practice not against the law and regarded by both prosecutors and defendants as a measure to protect their respective cases against jury tampering. The Supreme Court did not agree with the trial judge that Sinclair's action obstructed justice, but it sustained the judge on the ground that it had "a reasonable tendency" toward the obstruction of justice, and upheld the lower court's sentence of six months in jail. The two contempt sentences were made concurrent.

6. Doheny was acquitted in a jury trial of criminal charges to defraud the government by conspiracy.

7. Ex-Secretary of the Interior Fall, described by the Supreme Court in the civil case decisions as "a faithless public servant," was convicted in a jury trial of criminal charges to defraud the government and sentenced to one year in jail and a $100,000 fine.

No wonder that the Raleigh, North Carolina *News and Observer* declared that " 'Who bribed Fall?' may be absorbed into the folklore of a growing Americana to take its place beside the historic query, 'Who killed Cock Robin?' "

The fact that these decisions were protracted over six years was not the only reason for confusion. During this period, the Naval Reserve scandal took second place to a matter that had no legal connection with it. It was described by Paul Giddens, a Standard Oil of Indiana historian, as "one of the largest and, at the same time, one of the most unusual and mysterious oil deals in history."

In preparing their civil suit against Mammoth Oil, the

government prosecutors investigated Fall's bank accounts and found references to Liberty bonds. These were traced, by their numbers, to the Continental Trading Company of Canada, which had purchased $3,080,000 worth. The company had been organized in 1921 to buy Mexia oil from Colonel Humphreys (the Texas field's discoverer) and sell it to Sinclair Crude Oil Purchasing Company and Prairie Oil and Gas Company. The company was dissolved in 1923.

The government was unable to find out how the Continental Trading Company Liberty bonds came into Fall's possession. Neither he nor Sinclair, both facing criminal charges, would testify.

Col. Robert Stewart, president of Standard Oil of Indiana (joint owner of Sinclair Crude Oil Purchasing Company), left the country on business when the civil trial went to court. James O'Neill, president of Prairie Oil and Gas, resigned and left the country. Henry M. Blackmer, organizer of Continental Trading Company, left the country. The Canadian lawyer who had handled Continental's business refused to reveal who his clients had been.

The public immediately concluded that the Standard Oil companies were linked to the oil reserve scandals. In 1928 the Senate decided to delve into the matter of the bonds.

The first witness before the investigating committee was M. T. Everhart, Fall's son-in-law, who testified that $233,000 in Liberty bonds were delivered to him in May, 1922, in Sinclair's private railroad car, as he and Fall sold a one-third interest in their Tres Ritos Cattle and Land Company to Sinclair. Of the bonds delivered, $200,000 worth had been purchased by the Continental Trading Company. Fall and Everhart used part of the bonds to pay their company's debts; the rest were deposited in Colorado, Texas, and New Mexico banks. These were the bonds whose origin the government had tried to trace.

This sensation was followed by another when Will H. Hays, former Republican National Committee chairman, testified that Sinclair, in November, 1923, had contributed $160,000 to the Republican National Committee, all of which was in Continental Trading Company Liberty bonds.

It was not until after Sinclair was acquitted in his criminal trial that both he and Colonel Stewart testified before the Senate committee and told the story of the Continental Trading Company.

Henry M. Blackmer, president of Midwest Refining Company—the Salt Creek, Wyoming, company that merged with Standard of Indiana—persuaded Colonel Humphreys to sell 30 million barrels from his Mexia discovery for $1.50 a barrel. Blackmer organized Continental Trading to sell the oil to Sinclair Crude Oil Purchasing and Prairie Oil and Gas for $1.75 a barrel.

When the contracts were being prepared, Colonel Humphreys said, "Gentlemen, I wish you would make this contract for 333,333,333⅓ barrels of oil because that will make the consideration fifty million dollars, and I have some pride in wanting to be a party to a contract of these dimensions."

Everybody concerned was pleased to accommodate Colonel Humphrey's pride. This meant that Continental Trading would receive $8 million in commission. The oil was to be paid for monthly as produced. At $1.75 a barrel it was still under the market price. Sinclair discussed the commission privately with Blackmer, saying he would expect his company to get a fair share. Blackmer agreed, and later Continental's attorney advised Colonel Stewart he would receive a part of the commission.

Continental Trading collected its commission each month, turned it into Liberty bonds, and divided them equally between Blackmer, Sinclair, Colonel Stewart, and James O'Neill. In May, 1923, Continental Trading sold its contract to Sinclair Crude Oil Purchasing and Prairie Oil and Gas for $400,000, having distributed approximately $4 million in rebates.

Stewart, Sinclair, and O'Neill had not advised their companies they were receiving a rebate. Colonel Stewart told Standard of Indiana's lawyer confidentially about the deal and made him a private trustee of the bonds, which were kept in a safe-deposit box with instructions that they belonged to either Standard of Indiana or Sinclair Crude Oil Purchasing. When the matter became public, Colonel Stewart turned the bonds over to his company. O'Neill did not return to the United States, but his son returned his bonds to Prairie, saying his father had always considered them Prairie's property.

Sinclair also turned over his bonds to his company, saying he had been holding them for the company until all the trials were finished. As to why he had used the bonds to buy an interest in Fall's ranch and to contribute to the Republican National Committee, he had what to him was a perfectly logical reason:

"I had the bonds put in a vault along with several million dollars' worth of other Liberty bonds of the same issue which I bought," he said. "There was no reason to keep them separate. It was like putting a lot of thousand-dollar bills with other thousand-dollar bills. From this lot of bonds I paid for my share of the ranch."

Incomprehensible as this might have been to a public unaccustomed to putting lots of thousand-dollar bills with other thousand-dollar bills, much less several million dollars' worth of Liberty bonds together, no one who knew Sinclair, and particularly the officials of his company, considered this anything other than his normal behavior. As one company official described him, "Above all, he was MacGregor! Where he sat, there was the head of the table."

Harry Sinclair and his company were one and the same thing. He never felt the necessity of consulting anyone concerning the conduct of his business. Although Colonel Stewart acted similarly, Standard of Indiana was not his personal company. The Rockefeller family controlled only about 15 per cent of its stock, but felt that public confidence had been seriously shaken by the blistering editorial attacks accompanying the hearings. John D. Rockefeller, Jr., successfully led a stockholder's proxy fight ousting Colonel Stewart. The issue was so important that ninety-year-old John D. Rockefeller, Sr., issued one of his rare public statements, calling on Colonel Stewart to resign.

The effect of the civil and criminal trials and the related Senate hearings was devastating to both Sinclair and Doheny. Public interest and censure had centered on them rather than on Fall.

Doheny was popularly supposed to be next to Rockefeller and Ford in wealth. Although this was far from the truth, his manner of living gave this impression. The Kansas City *Star* referred to him as "the Pullman porters' delight. He never tips less than a dollar and probably has not heard the jingle of silver money in his pocket for years."

His Los Angeles estate featured a vast conservatory containing the largest collection of tropical palms in the world. He built a natatorium to rival the Roman baths. His yacht, named *Casiano* after his and Charles Canfield's famous Mexican gusher, had a 7,000-mile cruising range.

When Villa ravaged Mexico in 1914, leaving starving communities, Doheny filled a ship with $200,000 worth of food and sent it to Tampico. In Los Angeles, he built a $1,500,000 Catholic church, St. Vincent's, and gave a $100,000 radium annex to the Episcopal Good Samaritan Hospital.

When he made a $25-million loan to the President of Mexico, he considered it good business to help the country from which he had received so much. When he bought $250,000 worth of Irish Republic bonds, he expressed the sympathies of his Irish heart. He was dismayed when the New York *Evening Post* charged him with a "mania for power. Like a medieval banker-prince, Doheny makes war on Great Britain in Ireland. He juggles the Mexican Republic in one hand and the Irish Republic in the other."

As Doheny told his life story to the Senate, he called himself "just an ordinary, old time, impulsive, irresponsible, improvident sort of a prospector."

"Just about as improvident as Calvin Coolidge," sniffed Washington newsman Raymond Clapper.

Edward Doheny broke down on the witness stand. It was a pathetic sight. The emotional strain of his voyage into the past, contrasted with his reason for being there, had been too great. How could he expect understanding from these men sitting in judgment on him, who did not know the meaning of dogged years of failure in the mines, or the excitement with which he and Canfield had swung their picks, digging for oil in Los Angeles, or how a man could forget all else in the world in the joy of discovering a bubbling oil spring in Mexico?

Doheny remembered how the country reacted to Charles Canfield's death in 1913. Canfield had lavished his money on others but never on himself. He established homes for orphans, gave to every charity, and no one knew how many thousands of dollars he gave to prospectors. But there was something greater in Charles Canfield's influence. No other wealthy man's death had ever brought forth such remarkable tributes from magazines and newspapers.

Harper's Weekly called him "the ideal American. All his life long in business and out of it, Canfield stood as an exemplar of integrity, energy, cleanliness and courage."

"He was the best type of American pioneer," stated *Collier's Weekly*, "literally clean in business and personal habit, who left no human wrecks along his road to success."

"Charles Canfield was one of God's noblemen," mourned the Los Angeles *Express*.

"A forceful agency for business integrity and honest building was lost to the country when he died," said the *Review of Reviews*. "He was a mighty builder whose bricks were sound, his mortar clean and whose hand was always out helping."

Edward Doheny had lost his balance wheel when his partner died. Although the soundness of his own bricks and the cleanness of his mortar had been questioned, the jury which tried him did not believe it. "A human factor played its part in the criminal case," the New York *Herald Tribune* commented. "That was the personality of Mr. Doheny grown old in the oil business, with a long record of frontier integrity. Whatever the jury felt with regard to Mr. Fall it clearly wished to believe the best of Mr. Doheny."

No acquittal could erase from Doheny's mind the public scorn and vilification he endured. Although Doheny the man lived until 1933, Edward Doheny the oil explorer and oilman died in the course of the Naval Reserve trials. While they were going on, he sold control of his world holdings to Standard of Indiana for $37,575,000. Included were Doheny's rich Mexican properties, Venezuelan properties producing 45,000 barrels of oil a day, an interest in concessions in Iraq, one of the world's largest tanker fleets, refineries in Mexico, Venezuela, and Louisiana, and marketing facilities along the Atlantic seaboard and in Europe and South America.

Acquisition of this property made Standard of Indiana second only to Jersey Standard as a world oil power.

Doheny retained his California properties and a stock interest in his Pan American Company, whose operations Standard of Indiana directed, but his creative days were over. He had learned that empire building and millions didn't matter half so much as remaining the ordinary, old-time prospector he had tried to represent himself as being, but which only his partner, Charles Canfield, had remained to the end.

Harry Sinclair bitterly served his contempt sentences. When he left the Washington, D.C., jail on November 21, 1929, he issued a statement.

"I was railroaded to jail in violation of common sense and common fairness," he said. "I was a victim of political campaigns to elect honest Democrats by proving how dishonest Republicans were. When I was assailed in this worthy enterprise, the Republicans discreetly replied that guilt, if any, was personal. I was politically assailed but not politically defended. The newspapers wept many tears over my defiance of law and my 'unchastened recalcitrance.' I cannot be contrite for sins which I know I have never committed, nor can I pretend to be ashamed of conduct which I know to have been upright."

Coldly furious, Sinclair was more determined than ever to demonstrate his power and capacities. The stock market had crashed, and Sinclair quickly saw that cash would be his company's greatest asset. In a panic, people rush to sell the future for the present; the man with cash can buy. He began negotiating with Standard of Indiana to sell them his half-interest in their jointly owned companies. Six months after his release from jail he consummated the largest oil cash transaction ever made, receiving a check from Standard of Indiana for $72,500,000.

This was the emperor turning defeat into victory. The cheapest way to find oil in these difficult days was to buy it, and Sinclair was ready. As the depression deepened, even his old Standard enemy, Prairie Oil and Gas, weakened. Sinclair acquired its production, pipelines, and markets.

Had there been no Teapot Dome, Sinclair would have regarded this as one of his greatest triumphs. But the golden dream of being a greater oilman than John D. Rockefeller had been tarnished. Sinclair professed to have his old ambition, but in his heart he knew that he could no longer accomplish it. He avoided public appearances, conducting his business at night in his Park Avenue apartment. Still the leader, still MacGregor, he now operated behind the scenes.

When Sinclair retired in 1949 he turned over to his successors a $700-million company, the seventh largest in America, but he hadn't licked Standard Oil. There were four Standard companies ahead of him. When he died at eighty, in 1956, a great many people were surprised. He had become such a legend they thought he was already dead.

What Emerson wrote of Napoleon Bonaparte could have served as Harry Sinclair's obituary: "It was not Bonaparte's fault. It was the nature of things, the eternal law of man and of the world which baulked and ruined him; and the result in a million experiments, will be the same. Every experiment, by multitudes or by individuals, that has a sensual and selfish aim, will fail. As long as our civilization is essentially one of property, of fences, of exclusiveness, it will be mocked by delusions. Our riches will leave us sick: there will be bitterness in our laughter, and our wine will burn our mouth. Only that good profits which we can taste with all doors open, and which serves all men."

A touch of the wine that burned the most, perhaps, before Sinclair died, was the knowledge of what a great tempest had been brewed in such a small teapot. Instead of producing the 130 million barrels the Bureau of Mines predicted, Teapot Dome was a little field that depleted itself after producing 2 million barrels. And Harry Sinclair paid for this knowledge with more than $30 million cash lost in wells, pipelines, and legal expenses. He paid more heavily still with his reputation and his dreams.

6 : The petroleum graveyard of Texas

LOOKING GLOOMILY at the financial statements of his Transcontinental Oil Company, Mike Benedum was having a hard time reminding himself of the truth of his own statement, "A wildcatter can't quit." Losses were increasing at an alarming rate—1921, $3 million in the red; 1922, $1 million more.

Now, in June, 1923, losses were already approaching the million mark again.

Joe Trees had been right in resisting his partner's idea of a big integrated company. Benedum, too, should have realized that Standard men had been perfecting that particular trade for sixty years. He should have stuck to his wildcatting. He laughed ruefully at the thought. That's just what the losses on the sheets in front of him were—dry holes. Magic Mike Benedum's failures.

"Mr. Pickrell is here," his secretary announced.

The young Texan who entered the office had an air about him that Mike Benedum recognized instinctively. It was that particular self-confidence, springing from basic optimism and honed by hard work, that is the stamp of the real promoter. Mike Benedum greeted him cordially.

"Mr. Benedum, I sure appreciate your seeing me," Frank Pickrell said. "I've got an oil well down in Big Lake, Texas, and everybody tells me you're the one man who can help me solve my problem."

"If you have an oil well, Mr. Pickrell, I shouldn't think you'd have a problem. The real problem is getting it."

The young man laughed. "That's what I thought, but let me show you."

He unrolled a map and pointed to a dot in West Texas. "There's our well and we've got 431,360 acres of leases from the University of Texas all around it. You won't find any better land except in the middle of the Sahara Desert. Nothing but sand dunes, mesquite, cactus, and not a drop of water within fifty miles. But it's got oil!"

Benedum frowned speculatively. "I know this part of the country. They call it the petroleum graveyard of Texas. There are seventy thousand square miles in West Texas, and every wildcat in the area has been a dry hole except here at Westbrook, ninety miles north of your well. Someone drilled there two years ago and got a 10-barrel well. Now I want to hear about your well."

Pickrell told his story. It was a familiar one but it had its own fascinating variations.

An orphan at fifteen, Pickrell had jerked sodas in his uncle's candy store in El Paso, then worked for a bank, doing so well he was able to buy an interest in his uncle's store. Putting his profits into mining stocks, he found himself worth a quarter of a million dollars at age twenty-three. When World War I began, he enlisted. On his return, he lost money in the stock market as quickly as he had made it. Like so many other restless ex-soldiers, he headed for Ranger with the idea of making his fortune. He stopped at Big Lake to see Rupert Ricker, the captain under whom he had served as sergeant.

Ricker was also thinking about oil. A University of Texas law graduate, he had returned to his home town to practice law. Poverty and drought were all he found. Reading about events at Ranger, Ricker wondered why there couldn't also be oil in this wasteland. The University of Texas, which owned more than 2 million acres in the area, answered Ricker's inquiry by sending him a geological report written by Dr. John Udden. There were possibilities "for the accumulations of oil and gas," Dr. Udden said.

Ricker boldly applied to the university for oil leases on 431,360 acres spread through four counties around Big Lake. When Frank Pickrell stopped to see him, Ricker was in desperate need of $43,136 to pay the 10 cents an acre annual rental.

Pickrell decided that Big Lake would be his personal

Ranger. With an El Paso friend, Haymon Krupp, he raised the money to buy Ricker's leases and pay the rental. The state required a well to be drilled within eighteen months. The partners organized the Texon Oil and Land Company. They found they couldn't even give the stock away in Texas. Pickrell finally managed to sell enough in New York to drill just before the deadline.

"We managed to get a rig set up by eight o'clock at night," Pickrell told Benedum. "We were congratulating ourselves about slipping in under the wire when I suddenly remembered we had to get an affidavit from two disinterested witnesses to the spudding of the well. We were sunk. There wasn't time to go to town and back, and there wasn't a chance of finding anybody around that time of night except jackrabbits or coyotes. We didn't know if we could get an extension and we didn't have money to pay more rentals. We just had to have proof we had started on time to hold those leases.

"Well, we headed for town anyway and it was just like a miracle—there was a car coming along with two fellows in it. They went back to the well with us, and I guess they thought we were crazy when I climbed up to the derrick top with a rose in my hand. As the boys started to drill I sprinkled rose petals and christened the well Santa Rita.

"You see, Mr. Benedum, some of the New York investors were Catholics and they had this rose blessed and asked me to christen the well to invoke the aid of St. Rita. The funny thing about it is, she's the Saint of the Impossible.

"The well was really blessed, because we had all sorts of trouble getting a crew to stay out there, getting water— we had to bring our drinking water by train from San Angelo, seventy-five miles away. But we always managed to accomplish the impossible right up to the last minute when she started flowing a hundred barrels a day at 3,038 feet.

"Our trouble now is, nobody will believe we've really found a field until we drill a second well. And we don't have any money. That's why I've come to you, Mr. Benedum."

Benedum didn't speak for a moment. He was struggling with his wildcatter's instinct and an unaccustomed caution. The Transcontinental financial statement did not encourage an expensive gamble.

"You wonder why anyone hesitates after you've taken the big risk," he said at last. "I'll tell you why. This may be a great field, but since there never has been one in West Texas nobody can be sure. So many lost money at Ranger they're afraid to get burned again. Frankly, I've got problems of my own at the moment, but since you've come to me for help, I'm going to suggest what to do."

Benedum made a list of companies for Pickrell to see in quest of financing. After the Texan's departure, the picture of his lone well out in the sagebrush kept floating into Benedum's mind. He knew that the majority of geologists believed only a gopher had any valid reason for digging a hole in West Texas. Yet Frank Pickrell had found oil, and hadn't he, Mike Benedum, always made his biggest strikes in just such vast unexplored territory? He decided to send Levi Smith, one of his most experienced assistants, to see the Santa Rita well and make a study of the area.

Smith's report was that of the practical oil finder. It seemed to him that even though the Santa Rita was not a big well, perhaps it was on the edge of a big field.

West Texas was still a question mark in Benedum's mind when Frank Pickrell came back to see him three months later.

"Mr. Benedum, I've gone to see all those companies and I

can't get any of them interested," he reported. "Won't you please take it?"

This time Benedum consulted his partner. Joe Trees advised against the venture. "If all the major companies have turned it down, we'd better stay out, too," he said.

Trees had been right about Desdemona and he had been right in staying out of Transcontinental's troubles. However, major companies had guessed wrong on Illinois and on the Caddo field. And what about Colombia? Benedum realized that he had known all along he couldn't refuse a new challenge. The Santa Rita well was still producing. There must be more oil where that was coming from.

Benedum made a deal with Pickrell to take a 10,000-acre block around the well, guaranteeing to drill eight wells and giving Pickrell a one-fourth interest in the Big Lake Oil Company he organized to handle the prospect.

Benedum's partners in Transcontinental did not want to participate, feeling Transcontinental's losses could be salvaged only by drilling prospects in proven oil territory. But by now Benedum had convinced himself that all of West Texas was big oil country. In order to explore in other areas besides Big Lake he formed the Plymouth Oil Company as a holding company for the Big Lake company. He subscribed a fourth of Plymouth's million shares and the balance were taken by associates and investors. Even Joe Trees caught some of his partner's enthusiasm and subscribed 50,000 shares. Benedum made Walter Hallanan, tax commissioner of West Virginia, president, and Plymouth, with $1 million capital, began drilling Big Lake's wells.

The physical difficulties were almost as arduous as establishing a camp in the Colombian jungle. There were no housing facilities, supplies, or water. It was a costly, disagreeable operation. The first dry hole was disappointing but not discouraging. Psychologically, the agreement to drill eight wells made everyone feel that the program was just beginning.

Number 2 was a teaser, pumping 12 to 15 barrels a day. Number 3 was a dry hole. Number 4 was another teaser. Plymouth appeared to be as ill-fated as Transcontinental.

When the next four wells were dry holes, the million dollars was gone and everyone but Mike Benedum was ready to write off a colossal failure. The locations his geologists had selected were all south, west, and north of the Santa Rita well. There was only one direction left where oil might be. Doggedly determined, Benedum put up money to drill a well 200 feet east of Santa Rita—and the ninth well came roaring in for 5,000 barrels a day. The great wildcatter had done it again. That bright summer day in 1924 the petroleum graveyard of Texas came to life. Within the next thirty years the explorers would find almost 10 billion barrels of oil in its 70,000 square miles. There were more giant oil fields lying hidden under this wasteland than would be found in any other single area in the United States.

Geologists would eventually learn that this desert land had filled in a great Permian basin that some 230 million years before had been covered by a sea rich in marine life As the mountain ranges uplifted, the sea subsided and the North American continent emerged as we know it today. Oil formed and accumulated in this basin, making it one of the country's richest oil regions. Yet, oddly enough, on the surface it was still as desolate, arid, and unattractive to any form of life as it must have been when the land was formed.

Big Lake brought Mike Benedum back to life, too. There were no more teasers. As wells marched eastward across the flat land, they flowed more copiously until they came in for 18,000 barrels a day. There were 132 million barrels in the Big Lake field alone and Benedum's Plymouth leases covered almost all of it.

Big Lake was Texas' most sensational discovery since Spindletop. The 2 million acres belonging to the University of Texas made it one of the richest universities in the world. On Thanksgiving Day, 1958, a nationwide television audience saw the University of Texas and Texas A. & M. College express their gratitude, just before the kick-off of their traditional football battle, by dedicating a memorial on the Texas University campus. It was the old Santa Rita rig, which had been blessed with rose petals. It started a flow of more than $300 million into their endowment funds.

Mike Benedum proved what happens when a wildcatter doesn't quit, but he also proved to himself that success comes not from determination alone. The imagination to grasp a new concept must come first, then the courage to try it.

Benedum was thinking about these things when Bill Wrather, the geologist who brought Desdemona to him, came with another challenge:

"Mike, I've got an oil field for you. This is a new geological theory and I doubt if you'll find another geologist anywhere who will agree with it, but I'm convinced it's right."

Wrather spread out maps of the Mexia Fault zone. Eight oil pools had been developed along the major faults. He pointed to a fault almost 3 miles west and paralleling the Mexia fault.

"This is on Nigger Creek," Wrather said, "and it's the only one in the area that's clearly exposed on the surface.

However, the producing formations dip up as they go west, and most geologists think you can't find oil here. They believe the major fold at Mexia acts as a dam to shut off migration westward, and since the Mexia field is so prolific it trapped all the oil.

"Leon Pepperberg, a geologist friend of mine, has the unique idea that oil migrated over the top of the Mexia fault while it was forming and that when this parallel fault occurred it trapped oil between the two faults. I've gone out in the field with Leon and I'm convinced he's right when he says the accumulation at Nigger Creek equals the one at Mexia."

Wrather's thinking appealed to Benedum. Like himself, Wrather could believe in the reality of something not yet proved. And the Pepperberg-Wrather theory was attractive. Besides, Nigger Creek was only 3 miles from a proven field, a fact that spoke as loud as any scientific theory.

On July 8, 1926, Transcontinental's Nigger Creek wildcat was completed at 2,864 feet, flowing 2,800 barrels of oil a day. Two weeks later a second well flowed 4,000 barrels a day. Transcontinental was so eager to slake its thirst that within two months it was producing 22,085 barrels a day from thirty-three wells. Benedum's company was finally out of the red.

Magic Mike's touch had returned, but a wildcatter can no more afford to become cocky than he can afford to quit. The Nigger Creek success was marred by Transcontinental's failure to follow the trail Plymouth blazed in West Texas.

In partnership with Ohio Oil, Transcontinental, drilled four dry holes 30 miles southwest of Big Lake. Benedum was preparing to abandon the area when his assistant, Levi Smith, came to him, obviously upset.

"I'm on a spot, Mr. Benedum. When I took these leases for Transcontinental on the Pecos River, we leased two ranches, the Stiles and the Yates. Mr. Yates is a fine fellow. Even before we hit at Big Lake he would come to tell me he was sure there was oil under his ranch. He wanted us to lease it. When we did, I guaranteed we'd drill a well on it. We didn't write that in the lease—he said my word was good enough. Now Ohio has drilled the test wells on the Stiles ranch and they're finished. Mr. Yates wants to know why we haven't drilled on his ranch."

"Levi, if you gave your word we'd drill, you know I'll back you up," Benedum said. "It looks like a lost cause, and I know Ohio thinks so. Tell you what: find out if they'll drill another well with us. If they won't, I'll put up the money and you and I will drill the well."

Although Benedum was noted for his unorthodox reasons for wildcatting, it would never have occurred to a company that he would drill just to keep somebody else's word. Ohio, suspecting he either had special information about the area or a good hunch, agreed to drill another well.

Ira G. Yates's 20,000 acres were so arid that he counted on more than a hundred acres per single head of cattle. Such conditions combined with the low price of cattle made ranching a thankless business. When the Transcontinental wildcat began drilling in an arroyo near his ranch house, Yates was down to a few hundred dollars and regretting the day he came to West Texas to make his fortune.

About one o'clock in the morning of October 29, 1926, after the well had been drilling a week, one of the drillers came up to the ranch house and shouted for Yates.

"We're down a thousand feet and we've got a big well,

Mr. Yates," he said. "It looks like it'll do several thousand barrels."

There was a moment's silence. "Well, I'll be damned," Yates said, and turned over and went back to sleep.

Yates slept until his regular getting-up time, 5 A.M. At breakfast his family wished him a happy birthday.

"Let's drive over and look at my birthday present," Yates suggested to his wife and his son.

At the well the Yateses talked with the drilling crew and stood around looking at the evidences of oil. There really wasn't much to see. The crew had managed to shut in the well. Suddenly, there was a rumbling sound.

"We'd better get away from here," Yates said. "Something is coming out of there pretty soon."

From a vantage point half a mile away, the Yateses watched oil, gas, and rocks roar out of their well, smearing the hillside.

"I don't know how much oil is there," Ira Yates told his wife, "but I know we've got something."

The rural telephone lines began buzzing. In almost no time the flat land was covered with miniature dust storms converging on the Yates ranch, the owners of every automobile in the area racing to see what had happened.

Ira Yates remained cool. He drew his son away from the crowd. "John, we didn't expect to get oil at such a shallow depth. It won't last long—it never does when it's so shallow. Son, we'll sell enough of our royalty while the excitement's high to pay all the debts we owe, and then, if they still want to buy, we'll sell a little more."

To Yates's astonishment, buyers were still clamoring for a hearing at midnight. He had now sold enough interests to bring him $180,000, so he decided to gamble by keeping the rest for himself. His judgment, under the circumstances, was excellent. At that depth, nobody could have suspected a giant field. Even Mike Benedum was incredulous when a few months later, deepened to 1,150 feet, the well became a mighty gusher of 70,000 barrels a day.

The Yates field became the nation's oil marvel. Its gushers rivaled those of Mexico. Some flowed 200,000 barrels a day. This was one of the great oil fields of the world. By 1978, only one-third of its 2 billion barrels had been produced.

Never had there been such a lucrative proof of the maxim "It pays to keep your word." Mike Benedum was dazed by this most spectacular strike of his whole career. Transcontinental and Ohio owned half the field.

Now that there was no question as to Transcontinental's success, Mike Benedum decided the grandiose dream of an integrated company competing with Standard seemed rather absurd. A wildcatter should stick to wildcatting. Ohio wanted to buy the company, so the trader buried the empire builder in the petroleum graveyard of Texas. Benedum traded Transcontinental's holdings for $60 million worth of Ohio Oil Company stock.

However, Benedum the empire builder refused to remain completely buried. He was soon resurrected in an astounding form. Even Joe Trees shared his excitement this time.

Bertram Lee, a Peruvian railroad engineer, came to the partners' Pittsburgh office in 1927 with a concession covering almost a third of Peru from the eastern slopes of the Andes down to Western Brazil. The area was 100,000 square miles, as big as Florida and Mississippi. A concession on all its vast unexplored natural resources could be had for building a $15-million railroad from the Pacific over the Andes and colonizing the region.

Benedum couldn't resist the temptation. He sent an expedition, headed by L. G. Huntley, a Pittsburgh geologist, to make a report. Seven months later Huntley reported 150 miles of oil seepages, great fertile areas capable of supporting a tremendous population, rich timber lands, potential fortunes in iron and precious minerals.

Characteristically, Benedum began on a grand scale, sending Maj. William Rhodes Davis to Europe as his emissary. Mussolini promised the new colony a million Italians. General Primo de Rivera, dictator of Spain, promised 500,000 Spaniards. President von Hindenberg promised a million Germans. Benedum hired anthropologists to work out the best ratio of the various nationalities.

Experts from all over the world began to plan crops, study markets, arrange for supplies and ships. A new market for American trade was being established. President Hoover congratulated the oilman for the extent of his vision.

By late 1929, Benedum had spent $1,600,000. The Old Colony Trust of Boston was prepared to underwrite a $60-million bond issue. The colony was ready to move off the drawing boards.

Mike Benedum reckoned without the stock-market crash. Further, there was a revolution in Peru. A new empire was not to rise from such ashes. Later, as World War II loomed near, the thought of a German-Italian colony at the head-waters of the Amazon was not appealing. Looking back on his wonderful plans, Benedum was moved to comment, "There is such a thing as a man's being too successful for his own good."

7 : The glorious seven fat years

LIKE THE FULFILLMENT of a Biblical prophecy, the desert blossomed and the glowing sands became a pool. As West Texas' black springs gushed forth, the state leaped to first place among the oil producers, a position it has never since relinquished.

In fitting tribute, Spindletop came back to life and began producing more oil than it had at the beginning of the century.

The old hill had been forgotten as the explorers surged across the country after each new strike, crossing and recrossing their own trails like hounds after a particularly eccentric jackrabbit. The ruins of Big Hill's mad days of glory were a depressing sight. Grass grew over rusted joints of pipe and rotting, oil-stained timbers, remnants of the derricks that had crowded the hill. The creaking of a few pumps, draining off the last barrels, was a ghostly echo of the raucous din that had been music to the boomers' ears.

Occasionally, through the years, the hill saw small flurries of activity as venturesome individuals and companies poked one more hole on the edges of the salt dome, hoping to find oil on its flanks. Then Big Hill attracted to itself another deliverer, as impassioned in his belief as Patillo Higgins and Captain Lucas had been in theirs.

Marrs McLean, a law graduate turned oil seeker, had found a little oil on the flanks of a salt dome in Louisiana. The more he studied Spindletop, the greater was his conviction that its flanks, too, had trapped oil as the great salt plug pushed its way to the surface. Four dry holes and dwindling capital did not persuade him otherwise.

Walking home in Beaumont with his next-door neighbor, Frank Yount, McLean spoke of the last test he was about to drill. Yount, a successful wildcatter who had found deep flank production on two other salt domes, agreed to drill not one but two wells. McLean turned over his leases, retaining a royalty. The first well went into salt 700 feet lower than any of the other wells drilled to the south, proving they were on the dome's flank. Yount located his second well about a quarter of a mile southeast of Captain Lucas' discovery. On November 13, 1925, Spindletop's second discovery well began to flow 5,000 barrels a day at 2,515 feet.

There was no boom. Frank Yount controlled almost all the leases. The old field, producing from the top of the dome, was ringed with dry holes. Now there would be an outer ring of producers bringing up 60 million more barrels from the flanks of the most famous of all salt plugs.

Beaumont grew staidly rich. The orderly second development was nothing like the old days. Something of the contrast could be seen during the visit of Ignace Jan Paderewski. The great pianist told his host, Frank Yount, he would like to see one of his gushers. Yount didn't want to disillusion him, so he decided to recreate the past. He instructed his drillers to rig a special connection on a well, and as he and Paderewski drove toward the field a geyser of oil spouted over the derrick top.

The pianist was entranced. Listening to the roar, he mused, "I think this is the kind of music I like best."

Everyone agreed that Oklahoma had seen its best days. There had been so much drilling that obviously there were no more giant oil fields to be found. However, Cities Service proved otherwise with a series of sensational discoveries, in-

cluding the greatest field ever discovered in this oil-laden
state. The old champion was not willing to give up the title
to Texas without an exciting recovery in the last round.

The first discovery was the result of scientific calculation.
Cities Service's Oklahoma subsidiary, depending entirely on
oil for its income, either had to buy or find new reserves. At
the moment it had no money to buy other people's discoveries and had to finance its exploration program out of current
earnings. H. V. Foster, the company's president, assembled
his geological staff and presented the problem.

"What we have to find," he said, "is a large untested area
with good possibilities and low-cost leases."

This, of course, was asking for the ultimate, but it did
not faze the team that had made the first great geological
discovery, El Dorado. The geologists carefully reviewed all
their information and then settled on an area 50 miles south
of Cushing. A blank spot of 20 square miles seemed to meet
the requirements. Part of the former Seminole reservation, it
had long since passed into the hands of small farmers. They
were oil-conscious, of course, but as there was no production
in the area they would not demand high bonuses for leases.

Geologically, the area was controversial. What was known
of the general subsurface conditions indicated it was highly
faulted and broken up. It was generally believed that either
there had been no oil accumulation in this area or else it lay
so deep it would be too expensive to find and produce. As
they concentrated on the area, the Cities Service geologists
decided the faulting was an asset and had probably formed
numerous traps. On the surface they could see folds of an
anticlinal type and they could find no reason why this
evidence of structure should not repeat itself underground.

Early in 1926 a test well was started 2 miles northeast of

Seminole, a flag stop on the Rock Island line. Seminole's railroad station was an old passenger coach dumped by the side of the tracks.

When the cable tool rig pounded below 3,500 feet the drilling crew joked about putting up a sign "Hell or China." At 3,720 feet there was more than a show—on February 28 they brought up 95 barrels. A few days later, 31 feet deeper, the well was flowing 1,170 barrels a day. The geologists were not as elated as they had hoped to be. Their oil was coming from a limestone formation from which only one field in Oklahoma had ever yielded profitable production.

Another company, stimulated by Cities Service's discovery, drilled a test a quarter-mile to the east. It was completed for only 60 barrels a day, but it produced from the Wilcox sand, known as the richest sand in Oklahoma. The game of oil checkers began to look exciting. Another company made a location about a mile southwest and completed a Wilcox well for 8,000 barrels a day. There was no longer any doubt. Another giant had been found.

Cities Service had leased conservatively. Newcomers were not too late. Another boom began. When Cities Service brought in another wildcat for 5,500 barrels a day in the Wilcox sand at Bowlegs, a mile or so south of Seminole, there was no holding things. Seminole was bloated with thirty thousand boomers and their traditional leeches, and the spillover ran to Bowlegs. To those who had experienced Cushing and Drumright, it was like seeing an old movie again: the same clapboard joints, saloons, tent cities. The same miserable roads. The same mud in which teams of mules bogged down until it covered their backs. The same unbridled violence.

It might have been Drumright rather than Bowlegs the

night a tough braggart was cut to ribbons in a knife fight. Instead of seeking a doctor, he drunkenly toured the town pointing to the blood pouring from his wounds. Still boasting of his toughness, he fell dead.

Then, like a tornado, the violence passed. The fifteen-year period between Cushing and Seminole was more than a passage of time, it was a growing-up period. The old ways would no longer do. Cities Service, realizing the magnitude of its discovery, built a model camp in the Seminole area with residences, bunkhouses, mess hall, recreation hall, ice plant, garage, children's playgrounds, and tennis courts. Other companies followed suit. Anyone looking for trouble could always find it, but it would never again be accepted as a necessary evil.

Seminole provided as mad a drilling race as the state ever witnessed. Everyone drilled at once to get the oil before his neighbor got it. Cable tools were abandoned in the rush, now that oil had been found, and the field was thick with the faster rotary rigs.

One driller was frustrated at 2,000 feet when his pipe froze—his bit, caught fast on something and would not budge. The driller on another rig 660 feet away, was equally perplexed. His bit, 2,000 feet down, would go no further. To his astonishment, the mud coming up the drill pipe was filled with iron cuttings. They had each drilled crooked holes and they had converged, the drills locking together.

The Cities Service geologists were amazed at what their calculated approach had discovered. The 20-square-mile blank spot on the oil map contained five giant fields, totaling over 800 million barrels, and there were dozens of smaller oil fields. Cities Service discovered three of the five giants.

Within a year the Greater Seminole fields were producing 10 per cent of the United States' oil.

But the Seminole discoveries were only a prelude. In their scouring for prospects, the Cities Service geologists found evidence of structure near Oklahoma City, the center of the state. Despite the many dry holes drilled in the area, Cities Service had such confidence in its geologists that it leased a 6,000-acre block and, in June, 1928, began drilling about 6 miles south of the city.

In 1913, as the Cushing boom reached its height, a group of Oklahoma City merchants had formed the Merchants Oil Company, subscribed $15,000, and drilled a 3,000-foot well. They hadn't gone deep enough. Close by, on December 4, 1928, at 6,335 feet, the Cities Service well blew in, running wild, for almost 5,000 barrels a day from the Wilcox sand.

The celebration was none the less vigorous for being delayed—the cannon which Oklahoma City fired when the well drilled in saluted the discovery of Oklahoma's greatest field and the nation's eighteenth most productive, containing 750 million barrels. As it developed, stretching out for 11 miles, it spread through the city's east side. Some of the richest production lies under the state capitol. The tall trim derricks on the capitol grounds are the perfect landscaping touch for a state that owes so much to its subsoil wealth.

Although there was a frantic scramble for leases, and oilmen descended upon the city like a flock of crows on a cornfield, Oklahoma City marked the end of the boom-town era in Oklahoma. Even before the field reached the city limits, Cities Service hired a municipal planning engineer, placing his services at the disposal of the city government and Chamber of Commerce. An orderly development program was carried out for the expanding population.

There was one unscheduled development in the careful planning, a sensational event that made the Oklahoma City field world famous.

At daybreak one cold March morning in 1930 a tired crew was pulling the drill pipe out of a Cities Service well in the south end of the field on the Mary Sudik lease. The bit had penetrated the oil sand at 6,471 feet and the crew was preparing to set permanent pipe to complete the well in routine fashion. The men carelessly neglected to keep the hole full of mud, which would have sealed off the flow of oil and gas until they could set pipe.

Sleeping giants must be handled with care. As though retaliating for being disturbed, there was a roar and the giant spit out the tools still dangling in the hole. A great column of gas shook the derrick, tossing lengths of heavy drill pipe stacked at one side as though they were paper straws.

The frightened crew worked frantically to put a control over the hole, but the well was busy drilling itself in. The column of gas blew more strongly and it began to rain oil. A strong north wind sprayed nearby pastures, cows, and farmhouses. Crews from other wells rushed to help but the wild well defied taming. Firemen and city and state police sped to the Sudik farm. A fire zone was set up, and all drilling wells in the area were stopped. Traffic was rerouted on highways miles from the well, which roared on, belching 20,000 barrels of oil and 200 million cubic feet of gas a day.

The Wild Mary Sudik was the news sensation of the world. The thought of a big city held at bay by an uncontrollable, dangerous force of nature captivated the imagination of people everywhere. Floyd Gibbons, the famous war correspondent, broadcast twice daily from the well on a nationwide radio hookup, the roaring making a strange background for his staccato descriptions of the courageous men, in slickers and heavy goggles, steel helmeted and with cotton stuffed in their ears, futilely attempting to cap the mad hole. Engineers came from afar to advise, suggest, but nothing availed.

On the sixth day the wind shifted to the south and now Oklahoma City was blanketed with gas. The citizens were panic-stricken. On the tenth day there was a momentary victory. A special connection was swung into place over the damaged surface pipe, slowly lowered, and attached. The sudden quiet was almost more astonishing than the continual roaring.

Wild Mary Sudik refused to be so easily imprisoned. Within a matter of hours sand in the hole, whirled by the fantastic gas pressure, cut a dozen crevices through the connection. The well went wild again, but in an even more terrifying way. Before, the gas and oil blew straight up. Now it sprayed in every direction like water from a hose with a thumb over the end. The force knocked men down when they tried to approach. The spraying sand cut them like knives.

The superintendent of the American Iron and Machine Company (appropriately his name was H. M. Myracle) now tried a device he had been building. First, a special die collar to cut new threads in the surface pipe was swung over the hole. When this held, a new section of pipe was inserted with a special master gate to shut off the flow. This time, on the eleventh day, Wild Mary Sudik was brought under control.

Exhausted, nerve-wracked people, forbidden for miles around to strike a match even to cook their meals, began to clean up the countryside. Thousands of acres of oil-soaked land had to be plowed under. Hundreds of buildings had to be repainted. Almost a quarter of a million barrels of oil was skimmed from pits, ravines, ponds, and streams. How much oil and gas was lost nobody could estimate.

Mary Sudik, who, with her husband, owned the dairy farm on which the well was drilled, had gone into hiding. She was offered a vaudeville tour and a part in a movie, but the quiet farm wife wanted only to forget her famous namesake. Soon the Sudiks moved from their ravaged farm to a modest home in Oklahoma City, to enjoy oil in the less hectic form of royalty checks.

As the twenties drew to a close, it seemed incredible that they had begun with fear of an oil shortage. It had been a decade of amazing oil finding. Oil production tripled. In 1929 more than a billion barrels flowed from the wells. Forty-two giant oil fields were found in Texas, California, Oklahoma, New Mexico, Arkansas, Wyoming, and Louisiana. In ten years more giants were found than in the previous twenty years. Hundreds of smaller fields were discovered.

As geologists looked at the record, they saw that three-fourths of the discoveries had been due to their efforts. They had answered the question Charles Gould posed in 1912, "Why bother with a geologist at all?"

As the great flood was released from the earth, the major companies expanded refineries and built thousands of miles of new pipelines. Service stations sprang up at every crossroads and on every corner, for the flood of oil was matched only by the endless stream of cars rolling off assembly lines.

The twenties in America were the greatest boom period any nation had ever known, and the symbol of the time was the automobile. In 1929, 4,587,000 new passenger cars were produced, a record that would not be surpassed for twenty years. There now were more than 23 million cars and 3,-379,000 trucks and buses on the roads, four times as many as there had been in 1919.

As refineries poured out millions of gallons of gasoline to power them, they also met the phenomenal fuel demands of ships, airplanes, factories, tractors, and home furnaces.

America was on a joyous, giddy buying and building holiday. Everyone was gaily confident the millennium had arrived. During these golden years, no one was more certain it would last forever than Ernest W. Marland. He was walking the earth like a king in the company of bankers and brokers, wearing his crown with the ease of hereditary royalty and viewing the nation's prosperity with personal pride. Who had contributed more to it than he? Being the creator of new wealth, he was correct, was he not, in feeling superior to mere financiers? Where there had been nothing, he waved a magic scepter and oil gushed forth—from Ponca City, Burbank, Tonkawa, and Seal Beach, California. This latter find, made in 1926, was especially pleasing to him inasmuch as he had opened a 175-million-barrel field on leases that Shell Oil had surrendered after sinking four holes and finding nothing of commercial value.

In England Marland made a deal with the Hudson's Bay Company to explore for oil on their vast holdings in Canada. He became a partner with Frank Pickrell in the development of his share of the Big Lake field in West Texas. He was exploring for oil in Colorado, Wyoming, New Mexico, Kansas. Products from his big refinery in Ponca City were being marketed in every state and seventeen foreign countries. Five thousand tank cars were emblazoned with a big red triangle, the words "Marland Oils" across it. The triangle decorated more than six hundred service stations in the Midwestern states.

As Marland grew richer, so did everyone associated with him. His philosophy was to share. The magnificent salaries, bonuses, and opportunities of Marland employees were the envy of everyone who didn't work for him. His charities were more lavish than ever. When Marland bought a Mississippi plantation for $750,000 to use as a hunting base, he also made it a model living place for the families living on it, building them houses, a school, a church, a community cannery. He would not allow scrip in his stores, wanting his tenants to be free of their traditional economic slavery.

Marland built a new mansion in Ponca City, modeled after an English manor house, and filled it with European paintings and art objects. The sculptors Jo Davidson and Bryant Baker made statues for the grounds—likenesses of Marland's family, friends, a Ponca chief, and Belle Starr, the famous bandit queen.

In 1927 Marland was no more aware than the nation as a whole that the glorious seven fat years had run their cycle. When the oil company showed a loss for that year, he felt it was only temporary and blamed it, in part, on his partners—those bankers also walking like kings, J. P. Morgan and Company—whose loans he had been using to such advantage in the expansion of his activities.

His association with them, at first so flattering to his ego, was beginning to annoy him. Their control of his board of directors had increased. They were not oilmen and so they vetoed his plans to build pipelines and thus achieve economies in transporting his own oil. They insisted he sell to Standard. However, at his partners' suggestion, he sold $30 million worth of Marland 5 per cent bonds and was gratified when the issue was oversubscribed four or five times.

When his company showed another loss in 1928, Marland realized the situation was serious, but had no doubt as to his ability to weather whatever financial storm might be brewing. The price of oil fell from $2 a barrel to $1. The West Texas and Greater Seminole fields flowed more oil than refineries could handle. Storage of crude and products reached alarming proportions. Everyone in the industry sensed trouble. The overproduction situation would become worse at the end of the year when the great Oklahoma City field was discovered.

When Marland met with his executive committee in New York in May, 1928, he was shocked to find that the bankers had turned against him. They told him the company needed a firmer hand, a president who would not be influenced by his friendship for the men who had helped him build the company. They offered him the board chairmanship. When Marland dazedly made recommendations for the presidency, he learned the bankers had already selected a Texas Company vice-president, Dan Moran, a hard-boiled operating executive who was the antithesis of the generous, grandeur-minded, polo-playing dreamer and prospector.

The bankers made it clear that the offer of the board chairmanship with salary was a pension. Marland would not be permitted any voice in the company affairs. Furthermore, he was told he could not live in Ponca City as his presence there would interfere with Dan Moran's reorganization.

Marland angrily resigned, asking his officers and old friends, who wished to resign also, to stay. It was a futile expression of love for what he had built. Dan Moran fired them all anyway.

Bitterly, Marland watched his name painted out of all the red triangles and replaced with a new name, Conoco. J. P. Morgan had merged Marland Oil with the Continental Oil Company, a small marketing organization whose roots went back to the old Standard Oil Trust.

With the disappearance of his name from the red triangles, Marland disappeared from the oil business as a creator of new wealth. He tried to make a comeback, but failed. Most of his money was gone. About all he had left was his mansion. Even his reputation seemed to be gone. After two years as a widower, he married his adopted daughter, an act that irresistibly invited gossip.

Yet there was a day in April, 1930, when he basked in the old glory. At a party one night, before the debacle, someone had asked Marland why he didn't have a statue made of an Indian to commemorate the Vanishing American.

"The Indian is not the vanishing American," Marland said. "The pioneer woman is."

Delighted with his inspiration, Marland invited twelve sculptors to make models for the statue he had in mind. At a cost of $200,000, he then sent the models around the country, getting the public's reactions to them. Bryant Baker won with his conception of a young woman striding across the prairie with a bundle and a Bible in one hand, her little son's hand in the other.

When the magnificent bronze was ready for presentation to the state of Oklahoma, the governor declared a holiday. An elaborate unveiling ceremony was held at Ponca City. As Marland, in top hat and cutaway, listened to the crowd's roar of laughter when Will Rogers said he had come all the way from California to undress a woman, it seemed impossible that this moment was all that remained of the splendid past.

And it wasn't quite all. With the same spirit he had started his oil career, he now flung himself into politics. There was still enough glitter to his name, there were still enough grateful Oklahomans, to elect him to Congress. In 1934 he ran successfully for state governor. His salary was his only income, and it was characteristic of Marland to prefer the governorship, with a salary of $6,500, to being a Congressman at $8,400. Neither amount meant money as he knew it—he would not have minded losing either one on a single poker hand in the old days—but as governor he could be a leader again.

Marland was as ludicrous a figure on the rough-and-tumble Oklahoma political scene as he had been at Ponca City the day he arrived wearing Pittsburgh knickerbockers, Norfolk jacket, and spats. He naively sponsored a program of social reforms of such magnitude that it jolted the breath out of practical politicians and taxpayers alike. He tried to run the state government as he had run his oil company. He could not understand the violent opposition he aroused with imperial manners; his lavishness was as reasonable to him as his philanthropies had been.

The Brookings Institute, at his instigation, made the first scientific survey ever made of governmental problems in the state. He managed to get some of the reforms through. Homesteads were exempted from taxation, a Department of Public Safety was established with a highway patrol and drivers' license system, public schools were given state aid, and a Planning and Resources Board was organized. But the reforms that would have helped take the state government out of politics were as effectively quashed by the legislature, as were Marland's efforts to increase taxation to finance his projects.

Marland was too idealistic, too mannered, too much of a gentleman for the unpolished electorate, a great many of whom had once cast their gubernatorial votes for Al Jennings, the bank robber.

There was ominous talk of impeachment. Marland could no more understand why he had failed in this undertaking than he could understand why he had lost his oil empire. The whiplash of public abuse left wounds that never healed.

Oblivious to his faults as an administrator, Marland was also unable to recognize the corruption that swiftly spread around him. Sitting out his last nightmarish months in the governor's office, his one consolation was the familiar, comforting sound of oil wells being drilled outside the capitol.

As oilman he had failed, as governor he had failed. Nevertheless, he had achieved two successes while in the executive mansion that would assure him a place in state history and oil history.

The Oklahoma City field had eventually surrounded the capitol grounds. Marland knew that wells would drain away oil belonging to the state. His proposal to lease the state's land was vigorously opposed by an astonishing variety of people. Nearby residents didn't want their real estate marred by derricks. Public-spirited citizens insisted that drilling would spoil the beauty of the capitol. Much of the opposition, Marland suspected, was being promoted by companies that would profit by the drainage.

As a lover of beauty, Marland took second place to nobody. Soon after becoming governor, he had, at his own expense, landscaped the capitol grounds with a great quan-

tity of shrubbery from his Ponca City estate. In the present
case he saw no conflict between beauty and economics.

A battle developed between the city and the state as to
which had authority to permit drilling. When the city council
refused to include the state-owned land on a ballot for a
zoning election to give leases, Marland acted with impatient directness. He decreed the capitol grounds a military
zone, patrolled them with state militia, and announced that
the state would receive open bids for leases. Knox Garvin, a
state senator, made the high bid of $374,000 and a one-fourth
royalty. The following day, surrounded by guards and news-
papermen, Governor Marland drove the stake for the first
well near the executive mansion. He not only ignored in-
junctions to restrain the state from such action but appointed
Knox Garvin a sergeant in the National Guard and made him
state drilling superintendent in addition.

Marland's satisfaction over winning this battle and its subsequent lawsuits increased as he watched construction begin
on a much-needed legislative office building. The state's oil
royalties were paying for it.

This achievement was eclipsed by Marland's successful
sponsorship of the Interstate Oil Compact, uniting the producing states in exercising control over the conservation of
their petroleum resources by preventing physical waste in
any form. He had written this plan into his platform, and
even before officially taking office had invited the oil-state
governors to his Ponca City mansion to work out plans for
the Compact. There were many arguments, but Marland
skillfully brought about agreements from which grew an
organization supported by twenty-one states. Never again
would wasteful boom-field practices dissipate the nation's
oil resources. The participating states formulated and en-

forced regulations requiring good engineering practices and waste prevention from overproduction and overdrilling. There had been abortive attempts to form such a compact before, but it was Marland's impassioned efforts that finally made it a reality.

The derricks on the capitol grounds and the Interstate Oil Compact Commission strengthened Marland's conviction that he was a statesman. In 1936, although he still had two more years as a governor, he decided to run for the United States Senate. Defeated, he ran again in 1938. The humiliation of the second defeat was softened only by his sense of relief on ending his governorship.

He returned to Ponca City, but not to his mansion. No longer able to keep it up, he sold the great manor house, which had cost him several million dollars, to the Carmelite Friars for $66,000. Marland finally lost hope. Sick and broken, he died in 1941.

Perhaps the only thought left to comfort him was that he was leaving his young wife, Lydie, a small home and some income. Yet Lydie, too, added a poignant touch to the Marland story. After living obscurely in Ponca City for twelve years, one day in 1953 she disappeared, and neither her friends nor the police could find her. Just as unobtrusively, in the summer of 1975 she returned to Ponca City, where she continues to live quietly.

In 1948 the Marland mansion was sold to the Sisters of St. Felix, who added Angela Hall, a chapel, dormitory, and administration building. In 1975 the City of Ponca City, with help from Continental Oil Company, purchased the estate to preserve the heritage of the Marland era and opened it to the public. In 1976 the National Petroleum Industry Museum and Hall of Fame was located on the grounds of the estate.

E. W. Marland had laughed on that day of eager beginning when Chief White Eagle warned him he would make bad medicine for himself when he drilled the Ponca burial mound. He had indeed made bad medicine for himself, but not by finding oil. How Marland really felt about the swift rush of events that engulfed him was revealed one night when he was driving back to Ponca City during his campaign for governor.

As the car came to the top of a ridge he asked his companion to stop. The great refinery he had built and lost sparkled with lights as it processed the oil from fields he had discovered. Lights twinkled from thousands of homes spread around the refinery—homes built because he had found oil. He thought of the red triangles from which his name had been removed—triangles on service stations, tank cars and oil properties throughout the United States, Mexico and Canada. There were five thousand men working on those properties which were worth $125,000,000. Millions of Americans had driven their cars, heated their homes, flown their airplanes and powered their factories and businesses with the millions of barrels of oil that poured from his giant discoveries.

Marland sighed as he contemplated the twinkling lights. "Well, there it is. The work of my life. It must mean something. Anyway, they can't take that away from me."

8 : *The greatest gamble*

ONE SUNDAY MORNING in early September, 1930, Ed Laster drove slowly down a road in Rusk County, Texas. The calendar had changed months, but the weather had not. Only the trail of fluffy white cotton pods blown from trucks and

wagons going to the gin officially proved that the long hot
summer was over.

Laster's eyes brightened as he saw a young farmer walking
across a field. He stopped the car and hailed him. "I'm looking for some help," he explained. "We're drilling a well over
on the Daisy Bradford farm and I'm short a hand on my
crew this morning. We pay five dollars a day. How about
joining up with us?"

The farmer laughed. "Somebody had too much corn likker
last night, huh? Well, I'll give it a try. That's more than I'd
make pickin' in low cotton."

Ed Laster hesitated. Corn whisky hadn't been the trouble.
The missing crew member had departed for a better reason.
"You might have to wait a little while for your pay," Laster
admitted reluctantly.

"Oh, that's all right," the young farmer said. "This is Dad
Joiner's well, isn't it? I know the trouble he's been having.
Seems to me we all ought to give him a hand when we can.
If he finds oil around here it'll be our salvation."

Laster relaxed. He should have realized there wasn't anybody left in East Texas who didn't know Dad Joiner's problems or his stubborn conviction that he was going to find oil.

Dad had been promoting and drilling in Oklahoma ever
since Cushing was discovered in 1913. He was seventy now,
and Ed Laster admired him extravagantly for the optimism
and grit that enabled him to forget his seventeen years of dry
holes and concentrate cheerfully on whatever prospect he
was drilling, confident that this time he would hit.

Dad had started in the mercantile business in his native
Alabama. Later, in Tennessee, he had tried politics and
been elected to the state legislature. After that, a born prospector, he had gone to Ardmore, in Indian Territory.

He seemed to be hounded by bad luck. Drilling a well at Earlsboro, in Seminole country, he missed out on the Healdton field at his back door in Ardmore. Then he was forced to abandon his Earlsboro well at 150 feet because he ran out of money. Had he been able to go on down a few thousand feet he would have discovered one of the Seminole giants that Cities Service found thirteen years later. When he drilled a wildcat near Cement, Oklahoma, he got only a show, not knowing that this was a signpost to another giant field that someone else would discover there in due course.

It made Ed Laster fighting mad to hear Dad referred to condescendingly as "just a promoter." Just a promoter, indeed! Of course he promoted each new wildcat by selling interests to buy the leases and drill. Almost everybody got their start in the oil business that way. There was no difference between Dad Joiner and Mike Benedum or Harry Sinclair or any other oilman except that after you hit you were called an oilman instead of a promoter.

Of all Dad's promotions, this East Texas venture set a record: it was his worst luck yet. Dad had bought a few hundred acres of Rusk County leases from an Oklahoma City syndicate because they were cheap and it was an area without any oil. Looking over the rolling pine-covered hills, he was sure the earth would yield something more exciting than cotton, sweet potatoes, and corn. Accordingly, he leased an additional 5,000 acres, setting the impoverished farmers to dreaming of wealth with his announcement that he was going to drill.

Dad had timbers cut for his derrick's foundation but they lay unassembled on the ground for a year or two. The geologists scoffed at the idea of finding oil in this part of Texas. There was no evidence of any structure, either on the surface or from the subsurface information of wells drilled in the general area. Dad found money raising particularly difficult but finally sold enough interests to start drilling in August, 1927.

One geologist, Dr. A. D. Lloyd, shared Dad's belief in the region's possibilities and recommended a location, but Mrs. Daisy Bradford, a widow, refused to lease her 975-acre farm unless a well was drilled on it, rather than just off. Therefore, Dad started the No. 1 Daisy Bradford two miles southeast of Dr. Lloyd's site.

His secondhand equipment was dilapidated. The drilling pipe was worn out. At 1,098 feet the bit jammed and could not be dislodged. Nothing to do but skid the rig a hundred feet away and start No. 2 Daisy Bradford. But the new hole was months getting started. Dad was having to sell $25 certificates for a small interest in the well. This time he got the hole down to 2,000 feet. Then disaster struck again: the pipe was hung in the hole. Dad made a trip to Houston to get some special equipment and find a new driller. He met Ed Laster. An experienced driller, Laster had just finished some successful gas wells. Dad explained his troubles. Laster agreed to work for him for part cash and part lease interests.

When Laster saw the No. 2 Daisy Bradford he knew the situation was impossible.

"That pipe's hung in there in such a way we'll never be able to pull it out," he told Dad. "You'll have to start a new well."

Dad groaned. "That's more impossible than you think getting the pipe out is. I'll never be able to get folks to put up money for a third well."

Weary and dejected, Dad was not defeated. An inner voice seemed to urge him on, seemed to promise eventual

success. He started on his rounds again with his $25 certif-
icates.

Dad was spruce and cheerful. His small wiry frame and
clear eyes belied his years. Over a cup of coffee in a diner
he would tell a waitress about his well. When he finished
she would be eager to buy a certificate entitling her to a
$25/75,000$ interest in the well and a $1/500$ interest in the syndicate that owned 500 acres around the well. He sold police-
men, post office clerks, storekeepers, railroad workers—anyone he could strike up a conversation with.

In May, 1929, the No. 3 Daisy Bradford started drilling.
Dad just couldn't talk to enough people to keep the hole
going down steadily. Some months the rig worked only on
Sundays so that prospective investors from Dallas and other
towns would find it operating.

Dad's pathetic efforts became a joke among the oil companies, particularly after one scout made twenty trips to the
well and never once found it drilling. But the No. 3 Daisy
Bradford was no joke to the farmers and the townsfolk
nearby. This was no longer just Dad Joiner's well. It was
their well—literally, in the sense of the $25 certificates they
all bought, more humanly in the way trouble unites a community and brings forth its vital reserves.

When the drilling crew was short, farmers would quit their
plows and lend a hand for a day or two. Walter D. Tucker,
a banker in Overton, 10 miles from the well, would frequently drive over after the bank closed, put on overalls, and
go to work. Mrs. Tucker cooked for the crew. When Dad
couldn't pay his grocery bill, the Overton grocer accepted
scrip against future production and let Dad use the back of
his store as an office. Dad issued so much scrip for supplies
and services that storekeepers and customers alike circulated

it as real money. Sometimes a farmer would lose patience and refuse to cut or haul wood for the boilers, but there was always another one who would.

Dad's lease on Mrs. Bradford's farm expired and for a dizzying moment he thought his three years' toil wasted, but Mrs. Bradford was nursing the well, too, and she gave him an extension. And then another.

Even when Dad felt he couldn't sell one more $25 certificate, the faith of these people rekindled his own. Their cotton failed in 1929 and again in 1930. In addition, everyone was struggling in the quicksand of the Great Depression. Dad didn't see how he could do it, but he knew he had to keep on for these people as well as for himself.

Ed Laster and the young farmer he had picked up were getting close to the well now. They could see a dozen old Fords and wagons drawn up at the side of the road. The usual Sunday crowd had started to collect. Nobody really expected anything to happen, but visiting the well was a communal habit, like going to town on Saturday.

"How deep is she now?" the young farmer asked.

"Just below thirty-five hundred," Laster replied. A year and four months to drill 3,500 feet, he thought. If we had a rig that wasn't held together with safety pins and bailing wire, and we had the money to pay the crew, we could have drilled this hole in six weeks.

"How much deeper 'fore you find the oil?" the farmer asked.

Laster laughed. "We're wondering about that, too. We're down to the point now where if it's there, we should find it."

Only a few days before, a friendly scout had sent one of

Dad's rock cores to a paleontologist in Dallas and another to a geologist in Chicago, asking for their opinions.

The report from Dallas had been negative. They could not expect to reach their objective, the Woodbine sand, which was the producing formation in the Mexia fields, for another 1,600 feet, but in this area the Woodbine sand would not contain oil. Dad shrugged his shoulders and told Ed Laster to keep drilling until they received the Chicago report. Then Dad went to Dallas to dig up new buyers.

Laster put the young farmer to work and tried to build up pressure in the dyspeptic old boilers that didn't seem to care for their diet of green wood.

At noon, the farm women spread out picnic lunches. The hot afternoon dragged on. The farmers had their visit out and the crowd thinned. At four thirty Laster decided to take another core before shutting down.

The crew began pulling pipe out of the hole, unscrewing the 20-foot joints and stacking them at the side of the derrick. It was hot, wearisome work. When the last length was out, the drilling bit was replaced with a core barrel, a long steel tube with sharp cutting teeth around its bottom edge. The long, monotonous trip back down the hole began, each joint of pipe being swung over the hole, screwed onto the joint below, and slowly lowered.

At last they were ready to core. The drill pipe began to rotate, and 3,536 feet down in the earth the sharp teeth of the core barrel started cutting a ring of rock. As it cut down, the core of rock pushed up inside the core barrel. When it was filled, clamps gripped the bottom of the core and the sweating crew began the second long trip out of the hole.

When the core barrel emerged and deposited its cylinder

of rock and mud in the waiting buckets, Laster examined it cursorily. "Nothing new," he remarked, "except it looks like we've run into a little shale."

After the well was shut down, Laster tossed the core and core barrel into his car, drove the young farmer home, and went back to his boardinghouse. Wondering if the Chicago geologist's report was going to confirm the Dallas report, he got around to picking up the core barrel. His eyes flew open. There, in the bottom of the core barrel, was nine inches of oil-saturated sand. How could he have missed it? He smelled it, tasted it, rubbed it in his palm. He rushed to the phone and called Dad in Dallas.

"That's fine," Dad said. "Write me a report. It will be good selling for the letter I'm getting out tomorrow to send to some new prospects."

Laster could understand the old man's reaction. After three years of trying to get a hole down, that was all he could think about. He couldn't absorb quickly what this news meant.

A picture flashed across Laster's mind: the two buckets still sitting on the derrick floor. Some of the sand was in those buckets. If an oil company scout should stop by the well, everybody in East Texas would know they had found oil and Dad wouldn't be able to pick up some leases he wanted.

Laster rushed back to the well. His heart sank. There was only one bucket on the derrick floor. Someone had already been there.

Ed Laster was back early the next morning waiting for a crowd of oil scouts he expected to converge on the well. The first to arrive was the friend who had arranged for the reports.

"Sorry, Ed," he said, handing him a telegram. "Chicago confirms Dallas. You won't get the Woodbine for another sixteen hundred feet and when you get there it'll just be salt water."

Laster was puzzled. He didn't care about the report now, but obviously the news wasn't out yet. Who could have taken that other bucket?

It was late afternoon before another oil scout dropped by. He laughed. "I found your bait, Ed, but you didn't catch a fish."

"What do you mean?"

"You know what I mean—that oil you planted in the sludge bucket. That was the prettiest Woodbine oil sand I've ever seen come out of the Van pool. Ed, you wasted your time driving fifty miles to get that sample. It didn't fool me a bit. I guess Dad's finally trying to unload, isn't he? Well, you can't use us to sell your stock."

Laster's quick resentment at the accusation subsided as he realized the scout had outwitted himself. The scout did such a job of convincing his associates that the oil sand had been planted, the major companies made no move to acquire leases even after Laster exhibited another core and the whole community was talking excitedly about nothing else.

When the tired old rig gave out, Dad was able to borrow another rig on credit. It didn't matter that the well had not yet been drilled in and no one knew how big a discovery it was. Everyone for miles around had waited so long for a miracle that they accepted its arrival on the slimmest proof. An oil boom without oil started.

Farmers, townspeople, and laborers began trading leases and royalties. As the word spread, others hurried to Rusk County. Within twelve days after the first showing of oil more than two thousand land transactions were recorded. Leases were selling for $10 an acre 15 miles from the well. The town of Joinerville was laid out on the highway where the road turned off to the well. Overton staged a "Joiner Jubilee," with street streamers praising Dad, a carnival, barbecue, fiddlers and dancing. Admittance to the barbecue grounds was by a badge carrying Dad Joiner's name.

Next, there were rumors that lawsuits would be filed against Dad because of the considerable confusion concerning the various syndicates he had formed and the thousands of small interests. It was a bewildering turn of events for the old man—jubilees and lawsuits over a well that still had not been completed.

The Tyler *Courier Times* spluttered indignantly, "Is he to be a second Moses to be led to the promised land, permitted to gaze upon its 'milk and honey' and then denied the privilege of entering by a crowd of slick lawyers who sit back in palatial offices cooling their heels and waiting while old 'Dad' worked in the slime, muck and mire of slush pits and sweated blood over his antiquated rig down in the pines near Henderson?"

Party lines, rural mail carriers, and storekeepers passed the word that Dad would complete No. 3 Daisy Bradford on October 1st. Pipe was set in a cement plug at the bottom of the hole. All that Dad had to do was drill the plug.

By mid-morning on the first, eight thousand men, women, and children swarmed around the well. Roads were choked with dusty old cars, buggies, wagons. Mrs. Bradford's young nephew sold a hundred cases of soft drinks. Candy, popcorn, and balloon vendors had their biggest bonanza since the county fair.

On the rig floor, sprightly, cheerful, waving to friends,

creditors, investors, Dad watched the drilling operations. When the cement plug was drilled through, a hush fell. The crew ran a bailer down the hole to bail out the mud. There was no need for the crowd to be quiet. Hour after hour the bailer went in and out of the hole, bringing out mud, but there was no sign of oil.

At sunset some of the people went home for supper. They soon came back. At midnight most of the crowd was still there. A great many bedded down in their cars and wagons. Those who went home returned the next morning to find that nothing had happened through the night.

All through the second day the crew continued bailing mud. Both Dad and Laster were showing the strain. On the third morning, when all the mud was out and still no oil flowed, Dad asked slowly, "What do you think, Ed?"

"We'll start swabbing," Laster said with false cheerfulness. "She's just being contrary." Laster could not bring himself to think that it might be a dry hole after all, that the show of oil had been just that—a show and nothing more. Countless other wells had never produced more than a show. Surely, Laster thought, this wouldn't be another.

The crowd, unaware that such a thing could and did happen every day, was still hopeful, still expectant. The change from bailing to swabbing created an interesting diversion. They watched the crew lower the swab, a curious contraption of steel and rubber, with valves that fitted tightly inside the casing. Its rapid lifting at the bottom of the hole created suction, pulling fine sand and mud into the well. Perhaps the action would open the pores of the oil sand.

The swabbing proved to be as monotonous a spectacle as the bailing. Then, late in the afternoon, a murmur ran through the crowd. Those closest to the well said they could see some oil in the mud brought out by the swab. "The fluid is rising in the hole," Laster told Dad.

There was a gurgling in the pipe. Laster barked orders to douse the boiler fire. Thousands of people silently watched a stream of oil begin to spurt from the casing. It gathered volume and oil and gas sprayed over the top of the derrick.

The crowd shouted and laughed and wept. Men tossed their hats in the air. One man gleefully ran under the spray and took a shower bath. Children painted their faces with oil.

Dad Joiner, pale and shaking, was engulfed by men and women hugging him, shaking his hand, slapping him on the back. "I always dreamed it," he murmured, "but I never believed it."

Ed Laster was as worried about Dad as he was about getting the well under control and flowing into the tanks, with the seething crowd dangerously hindering the crew. "You go on back to Dallas, Dad, and rest," he urged. "We'll find out how big a well we've got." Laster's estimate was 5,600 barrels a day.

The strain was too much for Dad. He was laid up in his Dallas hotel room two weeks later when a group of certificate owners, fearful they would not get their share of the profits, filed suit in Rusk County, asking that his properties be put into receivership. Dad was unable to attend the hearing, but the judge spoke for all East Texas when he said, "I believe that when it takes a man three and a half years to find a baby, he should be allowed to nurse it for a while. Hearing postponed indefinitely."

Ed Laster tried to picture to Dad what was happening on the Bradford farm, in Overton and Henderson and Joinerville. There was such a stream of cars on the highway leading to the well that the traffic could hardly move. Thousands of unemployed workers were swarming in. Derricks were going up in every direction from the well as fast as men could build them. Rusk County, East Texas, was pushing the clock back to the days of prosperity.

It all seemed unreal to Dad Joiner. The derricks weren't going up as fast as lawsuits were being filed against him in Dallas courts. The maze of financing through which he had wound his way to the completion of his well was so burdensome that he asked for voluntary receivership. With immense relief, a month after the receivership, the old man sold his well and the acreage surrounding it, and some 4,000 acres of undisputed leases, to H. L. Hunt and his associates in Dallas for $1,500,000.

Haralson L. Hunt had been a professional gambler in Arkansas. An oil lease, won in a poker game, had produced. After that, card-playing seemed a dull business. His gambler's instinct set Hunt on a course of oil exploration that eventually made him one of the richest men in America.

His prompt action in buying Dad Joiner's leases demonstrated why. Although Dad's discovery precipitated a boom in Rusk County, no one knew how big the field was and the major companies almost unanimously ignored it. The geologists who had said the oil wasn't there now said it wasn't much of a field, just flash production. Only a real gambler, facing such odds, would buy a snake-pit of lawsuits.

The day before Hunt bought Dad's properties, a wildcat was spudded on Mrs. Lou Della Crim's 900-acre farm at Kilgore, 12 miles north of Dad's discovery. The lessor was Ed Bateman, a Fort Worth newspaperman turned oilman. He financed his drilling much as Dad had, selling shares of stock to small investors all over the country and even in Canada. The news of Dad's discovery made his selling job easy.

On December 28, 1930, the No. 1 Lou Della Crim came in as an estimated 22,000-barrels-a-day producer. East Texas went wild. Geology or no geology, the area was obviously as full of oil fields as Greater Seminole. Within a few weeks the derricks at Kilgore were silhouetted against the sky as far as the eye could see.

The avalanche of humanity that rolled into Kilgore was out of all proportion to the size of the two strikes, but they were proof of new abundance at a time when the nation was sinking deeper into gloom and want. The strikes offered hope, opportunity.

The town of Longview, 12 miles north of Kilgore, was hungry to experience Kilgore's zooming prosperity. The Chamber of Commerce offered $10,000 for the first well in the town's trade area. Only twenty-eight days after Kilgore's discovery they paid off to J. A. Moncrief, his associates, and the Arkansas Fuel Oil Company, when they jointly drilled a huge producer 5 miles northeast of the town. The group gave the prize money to their two drillers.

Astonished oilmen and geologists began to count the miles on the map and study the evidence of the wells. The Longview discovery was 26 miles northeast of Dad Joiner's, with Kilgore in the middle. They were producing from the same sand. In a matter of weeks one huge producing well after another was completed in ever-widening circles around each discovery. It couldn't be, but it was—one great field instead of three.

Not an oil mind in the country was prepared for the magnitude of the field. The Oklahoma City field had been among the nation's largest in area, covering 20 square miles.

The fantastic East Texas field was soon producing over an area of 211 square miles. A forest of derricks stretched across East Texas, 43 miles long and from 3 to 12 miles wide.

The big companies moved swiftly with their checkbooks, paying millions for properties that, even after Dad's discovery, could have been bought for a few thousand. However, even after the field was a year old, the major companies still owned less than a fifth of the producing wells and proven acreage.

The four months between Dad's discovery and the Longview discovery, during which the major companies stayed out of the play, enabled thousands of small operators to get leases. East Texas was the genesis of more oil fortunes and independent oil companies than any other field in history. The field was so huge that even those who thought their leases were on the fringe or completely outside soon found they were in the fairway.

Among the first independents to arrive after Dad's discovery was Gus Glasscock, an East Texan. There were six Glasscock brothers. The older ones had started a small circus, and the younger ones had performed in it. The Glasscocks' high-wire act was a Ringling Brothers' 1911 headliner. World War I broke up the act. After the war Gus roughnecked in the Ranger field. Then he moved with the Texas booms. Eventually, he and his brother Lonnie were able to buy two drilling rigs.

Drilling for others and for themselves, they found a little oil and then dry holes. They moved their rigs to East Texas after Dad's discovery, buying what leases they could find, keeping one rig busy on contract drilling, using the other for themselves.

One of the first leases they bought was on a 4-acre Negro

school ground within the Kilgore city limits. While they were drilling, the schoolteacher, whose confidence they had won, brought a preacher around, explaining, "He preaches on Sunday and sells leases on Monday." The preacher said that many Negro farmers had been swindled out of their leases and that he was trying to look after his congregation and see that they did business with honest men. He had a lease with him he thought might interest Gus Glasscock. It certainly did. It was the lease adjoining the No. 1 Lou Della Crim discovery.

The Negro school well came in for 39,000 barrels and the Glasscocks also completed huge producers on the lease the preacher brought them.

More out of sentiment than confidence, the Glasscocks leased the farm near Longview on which they had been born. Some of the old circus cages still littered the barnyard. "This land ought to be good for something besides raising acrobats," Gus Glasscock said, "but nobody ever made an easy dollar out of it yet."

When their old homestead well came in for 20,000 barrels a day, no discovery of theirs ever meant so much. They paid the crew double wages back to the day the drilling started and bought new Stetson hats all around.

Drilling in the mammoth field was frenzied. During one week a new well was completed oftener than once an hour. Within ten months from its discovery East Texas brought chaos and disaster to itself and the industry.

When the field was discovered oil was selling for $1.10 a barrel. By June, 1931, with a thousand completed wells in the East Texas field producing 360,000 barrels a day, the price plummeted to 10 cents. Oklahoma City and Greater

Seminole were also in full production, and the nation, its industrial activities declining, could not absorb the deluge of oil.

One operator summed up the situation as he looked at a menu in an East Texas café. "Hell, I sell a barrel of oil for ten cents and a bowl of chili costs me fifteen."

The governments of Oklahoma and Texas attempted unsuccessfully to get oil companies and operators to prorate production. On August 4, 1931, Governor "Alfalfa" Bill Murray took action. A man who carried his lunch to his capitol office in a paper bag and conducted conferences in his stocking feet, he simply shut down twenty-nine Oklahoma fields by martial law "until we get dollar oil." He urged Governor Ross Sterling of Texas to do the same thing. Sterling sent the National Guard to East Texas to enforce the shutdown of wells that by this time were producing 848,000 barrels a day.

The price of oil crept up again and both states, in cooperation with the industry, were able to work out plans to restrict production on an equitable basis. Enforcing them was another matter. A great many operators, small and large, felt that the oil belonged to them and that nobody should regulate their activities. They devised more ingenious ways of running bootleg oil than the nation as a whole contrived to keep itself supplied with bootleg liquor.

During the day, many East Texas wells ran their allowable production into tanks where it was gauged by state inspectors. At night, with lookouts perched on the derrick tops, they ran bootleg oil through secret, buried lines to refineries or pipeline buyers who obligingly did not record the amounts.

An operator tripled his daily allowance from his one well

by erecting two additional derricks, pretending to drill, and passing the results off as producing wells.

One supreme individualist defiantly built a concrete block-house over his well controls.

Texas Railroad Commissioner Ernest O. Thompson, placed in charge of the East Texas field, had all the wells tested. The survey showed that the field could produce 100 million barrels a day, twenty-six times more than the entire world could consume in the same period.

The tests also showed that there was such gas waste from thousands of unnecessary wells drilled that unless the field was produced scientifically, with oil being withdrawn slowly, billions of barrels would be unrecoverable.

The operators began to realize that they were ruining, not bettering themselves, by flouting regulations that would stretch out the life of their discoveries. The new governor of Oklahoma, E. W. Marland, called his meeting of state heads and pushed through his Interstate Oil Compact Commission. Each member state enacted laws, and the men with guns were no longer necessary in the fields to enforce control.

Had conservation not gone into effect in the East Texas field, it would have finally yielded 1 billion barrels. It will now yield 6 billion barrels, with final recovery in the year 2030.

Just as East Texas was the last great boom, marking the end of riotous conditions in the fields, so it also ended the days of profligate waste. Though some of its first wells were drilled almost touching one another, under state regulations later discoveries would be drilled one to every 10 acres, 20 acres, 40 acres, or 80 acres. In gas fields, wells would be drilled one to every 640 acres. Oil operators would learn how to unitize their interests in a new discovery field, and by voluntary agreement produce it to their mutual advantage.

To geologists, "Remember East Texas" became a slogan capable of jolting them out of any tendency to become complacent. They had long known that the answer to the question of "How do you find oil?" was "Find a trap and drill it." Now they knew that the variety of traps was probably infinite.

The East Texas trap was formed on the western flanks of a great uplift extending through Louisiana and Texas. The waves of an ancient sea had lapped the edges of the uplift and a great sand bar formed along the beach. Gradually the sea covered the bar and the uplift itself; when it subsided millions of years later the bar became a thick layer of porous sandstone tilted against the uplift: a perfect trap for the vast quantity of oil which had been forming from marine life and now migrated up into it from the basin to the west.

The mantle of earth gave no evidence of this ancient activity, but even before Dad Joiner's discovery well there were geologists who, piecing together a few clues from wells scattered across the broad general region, guessed that something like this might have happened.

As early as 1915, F. Julius Fohs and James H. Gardner told some independents to drill near Kilgore. They in turn persuaded Roxana, the Shell company, to drill a test. However, Roxana's geologist-president, W. A. J. M. Van Waterschoot van der Gracht, decided to drill a mile and a quarter east of the recommended location. He thereby missed the field and drilled a dry hole.

While Dad Joiner was drilling his holes, Albert E. Oldham,

a young Amerada Petroleum Corporation geologist, was convinced oil must have been trapped somewhere in the area and recommended that his company buy leases scattered through an area 20 miles long and 8 miles wide. The executives, after estimating the cost at $150,000, said no. They were staggered when all the area Oldham sketched proved to be inside the East Texas field.

Humble Oil was one of the few companies holding leases in the field before Dad's discovery well was completed. G. Mose Knebel, a Humble geologist, thought of himself as a "red flag" geologist. He was always seeking the striking clue that would lead him to a new idea about an area, a technique that ultimately led him to the position of exploration manager of Jersey Standard. Knebel saw his red flags in the East Texas area and persuaded Humble to take some leases. He then pleaded with Humble to drill, but management was not convinced. When Dad's well came in Knebel was afraid the leases he had selected were too far away, although they proved to be in the field. "Of course, no one had any idea then of how big the field was," Knebel said, "because there never had been a field that big."

Another geologist, Jack Frost, of Dallas, also saw some of the same red flags that Knebel saw. Although he had no money to drill, he acquired some leases in the area and was the only geologist who personally profited by his belief in the area.

These imaginative thinkers were the exception. They were unable to change the skepticism of the majority until a tired old man, after three years of struggle, and without benefit of geological reasons, proved it to them all.

When Dad died in Dallas in 1947, the money from his great discovery had been exhausted by lawsuits and dry holes. He was living on money donated by sympathetic friends, among whom was Haralson L. Hunt. It came as a surprise to those who read Dad's obituary to learn that the full name of the discoverer of America's two greatest oil fields was Columbus Marion Joiner.

9 : *Crazy with dynamite*

A SMALL OIL FIELD discovered in Oklahoma in 1930 would ultimately mean more than even the treasure of East Texas, since the method of its discovery gave explorers a new tool to find billions of barrels of oil that were undiscoverable any other way. Everette Lee DeGolyer, the father of this new method, began work on it in 1927—the year Dad Joiner began drilling his No. 1 Daisy Bradford. Like Dad, DeGolyer had three years of failures; unlike Dad, when he began experimenting, he was already one of the world's most famous oil finders.

Son of a mining prospector turned farmer, Everette DeGolyer took up the study of geology at the University of Oklahoma in 1906. Learning that the United States Geological Survey encouraged students to go on summer field parties, he applied for the only job open with a party working in the Rockies—as cook. In later years he enjoyed attributing his success to macaroni and beer. His mother taught him to make macaroni and cheese so he could pass as a cook. The art of producing a cold bottle of beer on a field trip first brought him to the attention of Dr. C. Willard Hays, chief of the Survey, when he visited the camp on a hot day.

The following year Dr. Hays conferred with Sir Weetman

Pearson, who was discouraged by his Mexican Eagle Oil Company's lack of progress in finding oil. Pearson was one of the first to believe that science could help. Dr. Hays's oil thinking had changed considerably since the day he advised Captain Lucas there was no oil at Spindletop. He agreed to supervise the Mexican search and supply a staff. Remembering the alert young student of the summer before, he offered DeGolyer a job.

DeGolyer's geology professor tried to discourage him and his classmates from choosing petroleum as a career. "You boys were born too late for oil geology," he said. "All the big oil fields have been discovered—Spindletop, Glenn Pool, Illinois, California. There's no future in oil."

It was with a sense of adventure, rather than conviction, that DeGolyer accepted the Mexican offer. There was another factor. He was in love with Nell Goodrich, a young music teacher, and his salary of $150 a month, instead of the $90 he could make working for the United States Geological Survey on graduation, would hasten the wedding day. His eagerness to get married almost cost him his job. After a year of arduous and seemingly fruitless work in the Mexican jungles, he left one day for Oklahoma without telling anyone where or why he was going. He was afraid that if he asked for a leave to be married it wouldn't be granted. The oil could wait, but maybe the girl wouldn't. Nell Goodrich agreed to the marriage only when DeGolyer promised he would eventually get his college degree.

Returning to Mexico with his bride, DeGolyer found the atmosphere, if not the climate, decidedly chilly. In spite of the company's decision to overlook his unorthodox action, the Mexican adventure seemed to be over. Some of the wells had found oil, but not enough to justify the costly operations.

Sir Weetman Pearson, now Lord Cowdray, was about ready to pull out.

The year was 1910. A few miles south, even as DeGolyer was hacking his way through the jungle in search of a promising location, Charles Canfield and Edward Doheny were watching their Juan Casiano gusher blow in.

On Christmas Day, Mexican children excitedly broke open gay *piñatas* to see what treasures would spill out of these tropical versions of a Christmas stocking. Two days later, in the coastal jungle north of Vera Cruz, DeGolyer and Mexican Eagle broke open their own underground *piñata*. The well blew in wild, sending a column of oil as high as a 35-story building. For two months the great gusher spewed 110,000 barrels a day. Thousands of cattle died as a carpet of oil covered their grazing lands and contaminated all fresh water for miles around. Flowing out to sea, the oil coated the shores of the Gulf of Mexico for 300 miles. This flood was only a token. Potrero del Llano No. 4 eventually produced 130 million barrels, the greatest quantity any single well in the world has ever produced.

A few months after the discovery of this phenomenon, DeGolyer abruptly resigned to return to Oklahoma and finish getting his degree. Lord Cowdray, who now considered the young geologist his personal luck charm, ordered him back to work, arguing that a degree could only be a punctuation mark in such a promising career. When the astonished peer realized that he could not dissuade the determined young couple, he offered to pay for the educational detour if DeGolyer would promise to come back. Diploma in hand, DeGolyer became Mexican Eagle's chief geologist. He soon discovered another spectacular field, Los Naranjos, whose first well gushed 50,000 barrels a day.

DeGolyer's geological studies of Mexico gained him enormous prestige throughout the scientific world, and Lord Cowdray was hurt and angry when he resigned to do private consulting work, saying he could no longer confine his studies and energies to just one region. But Lord Cowdray understood his aspirations better than DeGolyer realized. In 1918 he called DeGolyer to London to advise on the sale of Mexican Eagle's properties to Royal Dutch-Shell. "Now, De," he said when the complicated negotiations were completed, "how would you like to have a million dollars to form a company and explore for oil just the way you think it ought to be done?" He laughed at the alacrity with which DeGolyer said yes. "I thought I'd finally ask you to do something you wouldn't say no to," he said.

DeGolyer suggested the name Amerada, since the company would operate in the United States and Canada. "Besides," he said, "it begins with an A, which will make it appear at the top of any list." With unbounded enthusiasm, DeGolyer organized his ideal exploration company, acquiring leases in Oklahoma and Texas. At the end of the first year he was at the end of his million dollars, and Amerada was at the bottom of any list that ranked companies according to their production. Amerada had drilled one well—a dry hole.

Anxiously, DeGolyer returned to London, wondering how to persuade the directors to give him another million. Lord Cowdray told him frankly that although they would hold a meeting to discuss his request, the other board members were so discouraged that they were not disposed to invest any more money.

As DeGolyer gloomily arrived at Lord Cowdray's office for the meeting, a secretary handed him a cable. It was from

John Lovejoy, his assistant in charge of operations in Oklahoma. WELL CAME IN LAST NIGHT FLOWING OVER TOP OF DERRICK was all it said. DeGolyer blessed and cursed in the same moment. Just before leaving, DeGolyer had given orders to start three wells, a last desperate hope that perhaps the miracle would occur. Now, which well had come in and how much was it flowing? In his eagerness Lovejoy had failed to transmit vital details.

DeGolyer waited until the directors were seated, then dramatically handed the cable to Lord Cowdray.

"Hmm. This was not arranged beforehand, I trust, De." Lord Cowdray read the cable to the board.

As questions poured out, DeGolyer confidently explained that the well referred to was at Phillipsburg, Oklahoma, was a Wilcox sand test and, of course, was producing between 1,500 and 2,000 barrels, that they owned a half-interest in 160 acres and could drill fifteen more wells.

Pencils flew over notepads. The postwar oil-shortage price of $4.50 a barrel times a potential of 16,000 barrels a day net to Amerada... Well now, Mr. DeGolyer felt he needed another million for the company's operations? Surely, that was too little to develop not only this discovery but the other splendid leases Mr. DeGolyer had acquired. Wouldn't it be sensible to put up four million dollars more working capital? Mr. DeGolyer offered no objections.

After the board meeting, Lord Cowdray drove DeGolyer back to his hotel. "De, just how sure are you that the well Lovejoy cabled you about is the Phillipsburg test?" he asked in an amused tone.

"I'm not a damned bit sure." DeGolyer grinned. He knew that Lord Cowdray approved the way he played his poor hand.

On his return to New York, DeGolyer learned that the crucial well was not the Phillipsburg test. It was a well in the Osage producing only 150 barrels a day. He conveniently postponed making a report until the Phillipsburg well was finished. The memory of Potrero del Llano No. 4 spurting its oil hundreds of feet in the air somehow seemed tame compared to the news that the Phillipsburg well was flowing 2,500 barrels a day.

Lord Cowdray's enduring conviction that DeGolyer was his personal lucky piece was borne out. At the end of the second year Amerada had twenty-one oil wells, three gas wells, and one dry hole—the dry hole being the first well drilled.

As Wallace Pratt admiringly commented, "The characteristic which most distinguished DeGolyer was intelligence—vivid, inquisitive, practical. No one who spent as much as an hour in his company could escape the amazing quality of his mind—his percipient, almost luminous intelligence."

Although the majority of other geologists, still in the first throes of their love affair with anticlines, were blind to the possibilities of other types of traps, DeGolyer was devising his startling new approach to oil exploration.

Spindletop had always fascinated DeGolyer with its oil accumulation around a salt dome. He interviewed Captain Lucas and Patillo Higgins to learn all they could tell him. Within three to five years after Spindletop's discovery, looking for topographic mounds like Spindletop with oil and gas seeps, explorers found and drilled about forty salt domes on the Gulf Coast. They found a few fields but most of their tests failed. Explorers and geologists alike began to lose interest; salt domes fell into greater disrepute during the next fifteen years when seven hundred wildcats drilled at a cost of $20 million located only six new domes.

Finding that American thought on salt domes had not progressed beyond the stage of bizarre theories, DeGolyer studied all the European researches and became convinced that salt domes were great plugs flowing upward from deep in the earth, and that as they pierced the rock beds they tilted them so that they dipped away in all directions, forming traps for oil on the sides or in the beds arched over the plugs.

DeGolyer was sure there were many undiscovered salt plugs on the Gulf Coast. The problem of how to discover them challenged him. While in England, he heard of experiments by a Hungarian physicist, Baron Roland Eötvös, with the torsion balance, an instrument used in laboratories since the eighteenth century to investigate and demonstrate the laws of gravitational attraction. Baron Eötvös devised a field instrument to search for irregularities in the earth's gravitational field. Since such irregularities were due to the unequal distribution of masses of rock of different specific gravities, DeGolyer reasoned that the instrument could be used to find salt plugs, as the presence of a salt plug lowers the pull of gravity.

He had a torsion balance built to his order and brought the first one into the United States in December, 1922, to test on Spindletop. He was elated when it outlined perfectly the known salt plug.

His pleasure was short-lived. Surveys of other known domes and areas where domes were suspected gave indefinite results. He did not know it then, but Spindletop was a freak, one of the few salt domes that could be detected by this method. Amerada drilled several prospects, all of which were dry holes.

After two years of failure, DeGolyer was about to abandon the method as too expensive and inexact when a survey of the Nash Ranch, in Fort Bend County, Texas, gave as definite and exact results as the Spindletop survey. Hopefully, DeGolyer ordered a test drilled. It discovered another freak, the Nash salt dome, and several million barrels of oil, the first oil field in the world to be discovered by a geophysical method.

DeGolyer was not enthusiastic about the success. "I think we should waste as little time as possible on our successes," he told his associates. "We should restudy critically and seriously the failures. That's the proper method to advance and perfect the technique of prospecting."

At the same time that he was experimenting with the torsion balance he also was trying another geophysical method, that of the seismograph. Responding to vibrations rather than gravity, seismographs had been used by the Chinese since earliest history to register earthquakes. Now the Germans were experimenting with the creation and recording of small artificial earthquakes, the result of dynamite blasts, in an attempt to determine underground formations and conditions. Again DeGolyer's inquisitive intelligence applied this thought to the search for salt domes and oil.

The method, known as refraction seismograph, measured the time it took sound waves from the dynamite explosion to travel through rocks to a recording instrument several miles away. Sound waves travel at different velocities through different types of rocks, traveling twice as fast through dense rock salt as through clays, shales, or sands. Theoretically, the quick reception of sound waves would indicate the presence of a salt plug.

In 1923 DeGolyer recommended that Lord Cowdray send a German crew, under the direction of the method's pioneer, Dr. L. Mintrop, to Mexico to make a refraction-seismograph survey. The attempt to find the southernmost extension of the famous Golden Lane fields was a failure.

E. W. Marland and Gulf hired Mintrop crews to work in Texas and Oklahoma, recording another year of failure, until at the end of 1924, the Gulf crew found a dome.

In two years of experimenting with both torsion balance and refraction seismograph, each had only one success to its credit and plenty of failures to "restudy critically." Meantime, the giant discoveries of the twenties were being made with no need of geophysical assistance.

Nevertheless, DeGolyer was as obsessed with the problem of salt domes in general as Captain Lucas had been with one in particular. Deciding that of the two methods, refraction seismograph offered the greatest possibilities for improvement as a technique, he found the best man to work on the problem, J. C. Karcher, a young University of Oklahoma physics graduate who was experimenting with seismograph at the Bureau of Standards in his spare time.

DeGolyer organized the Geophysical Research Corporation as an Amerada subsidiary, and gave Karcher a $300,000 research fund. Karcher fortunately generated ideas more quickly than his experimental crews could explode dynamite— A crew would use $18,000 worth of dynamite in a few days. Karcher's refinements of the refraction technique were immensely successful. In 1927 Geophysical Research set the astounding record of discovering eleven new salt domes in nine crew months in the Gulf Coast marshes.

There was a hectic rush among companies to develop their own techniques of refraction seismograph. Those who did not have crews would send scouts to follow the Geophysical Research crews, hoping to learn where they were getting favorable reactions.

"The game had glamour," DeGolyer said. "If a shooter on the crew got a fast reaction he would keep a poker face and then shoot blank territory in order to fool the scouts."

By the end of 1930 the entire Coastal area had been surveyed by refraction seismograph, some of it five or six times. The geophysical work cost approximately $20 million, but some forty new salt domes were discovered.

Although refraction seismograph was an overwhelming success, to the oil industry's astonishment, it wouldn't work any place except the Gulf Coast. This was because of the Gulf Coast's simple geological conditions. There were no other high-speed rocks, such as there were in other areas, to complicate the readings. The method could not give a clue to underground structures where the various layers of rock were uniformly hard or of the same speed.

DeGolyer was the first to realize an exploration principle which would be his chief guide in research. "A technique exhausts its usefulness by being used," he said. It was typical of the man that while the rest of the industry was still complacently searching for anticlines, and wondering if the new refraction seismograph method could do all that its users proclaimed, DeGolyer abandoned the technique and began looking for a new one. He urged Karcher to perfect a reflection seismograph, the project he had worked on unsuccessfully in his spare time at the Bureau of Standards.

The principle of reflection seismograph is the old echo game of the sound of a shout being reflected by a rock. The distance to the rock can be computed by timing the lapse between the shout and the echo, multiplying by a thousand (the number of feet sound travels per second), and dividing by two. Similarly, reflection seismograph generates sound waves by the explosion of dynamite near the surface and records the travel time to and from a buried reflecting rock bed in order to determine its depth. After shooting many points in an area, the depths are correlated; if the rock bed is arched or folded underground, the shape of the structure can be mapped to determine if it can possibly be an oil trap.

Simple as the principle is, developing a machine to gather the accurate information soon began to seem impossible. The glory that surrounded DeGolyer and Karcher after refraction seismograph's brilliant success faded as Amerada drilled successive dry holes on "structures" delineated by reflection seismograph. Virtually every other physicist in the field had condemned the project in advance as theoretically unsound, and when DeGolyer and his project were discussed in oil circles, he was referred to as "crazy with dynamite."

Even DeGolyer's board of directors argued with him that after three years of failures he had sufficient proof that reflection seismograph was valueless for oil prospecting. The continued heavy expenditures on the lost cause were unwarranted because the company was already finding oil without reflection seismograph.

DeGolyer sat at his desk in New York in 1930 studying three seismograph maps Karcher had sent him from Oklahoma. As a businessman, he knew he could no longer ride roughshod over the protests of his board. As a scientist, he knew he must have incontrovertible proof that reflection seismograph could discover an oil structure that could not be detected by any other means.

His eyes narrowed and he puffed thoughtfully on his cigar as he dangled one foot over the arm of his chair. "Well, this one's it," he said aloud. "It's really diagnostic."

DeGolyer began to chuckle, once the serious moment of decision passed. Nothing entertained him more than an unorthodox scientific action, except, perhaps, an unorthodox remark at any sort of gathering where others were being pompously conventional.

The diagnostic test he had decided to drill was in the center of the deepest surface hole in the Seminole plateau. Surface formations dipped into the hole from every direction. The reflection seismograph picture showed that far under the center of the hole the rock beds humped up into a little dome which might be a trap for oil. From surface indications such an idea was against all rational geology. Furthermore, the hole was ringed with dry holes.

On February 6, 1930, DeGolyer received news that the little dome was really there, 4,214 feet below the deep hole. When the drilling bit pierced it, the well, which opened the Edwards field, flowed 8,000 barrels of oil a day.

This was the most important well drilled in America since Spindletop: reflection seismograph revolutionized prospecting for oil as completely as Spindletop had done. Half of all the oil that would be discovered from that date on would be discovered by reflection seismograph in structures which would not have been found otherwise.

When DeGolyer's other two prospects found oil, all possibility of doubt was dispelled. Now it was the whole oil industry that went "crazy with dynamite."

There was a great wave of oil finding. The first seismograph crews going into an unexplored area could easily find the large structures whose presence had been unsuspected from available geological data. From 1930 to 1937 in Oklahoma alone, seismograph found 361 such structures, of which 146 produced oil.

But such phenomenal results were only temporary. Seismograph was still not the miracle: a direct method of finding oil. It could find concealed, deeply buried structures, but not all of them contained oil. However, seismograph reduced the odds. Only one out of twenty-four wells drilled for nontechnical reasons finds oil. One out of ten drilled on geological recommendations finds oil, while one out of six drilled on geology and geophysical recommendations finds oil.

As abruptly as he had resigned as chief geologist of Mexican Eagle, DeGolyer resigned as Amerada's chairman of the board in 1932. Essentially, it was for the same reason. In thirteen years he had made Amerada the largest independent company in America devoted solely to oil production. The "Father of petroleum geophysics" was only forty-six. "After all, you should have more than one child," he told his associates. "Besides, a desk is a ball and chain. It prevents man's progress."

DeGolyer had always been an independent thinker and his new challenge was to be an independent oilman. "To my mind, the basic and fundamental reason for the success of the oil industry is multiple individual effort," he said. "Multiple effort and many minds are as essential to finding oil as they are to solving the problems of science."

Moving to Dallas, he organized his own company and, in partnership with Karcher, who also left Amerada, organized a seismograph company, Geophysical Service, Inc. He was his own lucky piece now—it was not long before he was responsible for the discovery of five major oil fields in Texas and one in Illinois.

The Illinois discovery was the most dramatic proof in any area of the value of reflection seismograph as a new oil finding tool. After the Illinois boom, started by Mike Benedum in 1905, the surface clues were exhausted and drilling steadily declined until in the early 1930's it reached zero. Experts stated there was no more oil in Illinois and the state was scarcely mentioned in the oil trade journals.

On January 29, 1937, the citizens of Patoka, were stunned by the three words, in great, four-inch-high letters, on the front page of the Patoka *Register:* OIL BOOM HERE. To their amazement, a 1,500-barrel well, "the greatest strike ever made in Illinois" had been drilled on a seismograph structure found by DeGolyer.

As fast as seismograph crews could cover the ground, new structures were found, drilled, and produced. Within two years, approximately four thousand wells were drilled and the written-off area became the nation's fourth largest producer.

Prospecting did not monopolize DeGolyer's energies and talents. Taking a young partner, Lewis McNaughton, he opened a consulting office and made their services available to industry and governments. The firm immediately became the world authority on appraising oil properties. The United States Navy retained them as consultants on all its petroleum reserves. The Brazilian and Mexican governments retained them as advisors. DeGolyer no longer complained of any lack of challenges.

His discovery of the Old Ocean field in Texas is an excellent demonstration of the uncertainties, missed opportunities, and curious vicissitudes of oil exploration.

On the Gulf Coast, not far from Houston, DeGolyer's seismograph crew found a structure. Looking for a drilling partner, DeGolyer went to Wallace Pratt, of Humble. The two great geologists were fast friends, but they did not always agree.

"De, I'm sorry, but we have decided not to join you on this one," Pratt said. "You undoubtedly have found a structure, but we have just drilled three dry holes on three seismograph prospects you brought to us, so you will forgive me if I am dubious about this one."

To Pratt's chagrin, a few months later a 500-barrel well was completed on this structure at 8,651 feet, opening up the Old Ocean field.

DeGolyer enjoyed needling Pratt about "missing the boat on this ocean voyage," but it was soon his turn to be chagrined. After the discovery, he sold his half-interest in the field for $1 million to Harrison and Abercrombie, the partners he had found to drill it. Not long afterward they drilled deeper and found seven additional oil zones, from 9,800 to 10,600 feet. It was a giant oil field containing 200 million barrels of oil. The half-interest Harrison and Abercrombie bought from DeGolyer for a million dollars they now sold for $20 million; later sold the remaining half for $40 million.

The most chagrined explorer of them all, however, was Mike Benedum. Ten years before, when his Transcontinental Oil Company was at its lowest ebb, drilling one disastrous dry hole after another, he had leased almost all of the acreage covering the Old Ocean field. Drilling to 6,000 and finding no oil, he had given up his leases. And Benedum had made the same mistake once before. Had he deepened his dry hole in Oklahoma in 1907, he would have been the discoverer of the Cushing field.

As hundreds of new fields were discovered by seismograph, the 1930's proved to be an even more abundant oil-finding decade than the spectacular one just ended, but neither the industry nor DeGolyer yet realized the importance of the revolution he had started.

The nation, pulling itself out of the depression, was de-

pleting its oil resources at such a tremendous rate that it
annually consumed the equivalent of ten 100-million-barrel
pools. By 1937 Americans used in one year more oil than
was produced during the industry's first fifty years. Half the
supply came from seismograph-discovered pools.

If DeGolyer had not persevered in drilling that last diagnostic test in the deep hole of the Seminole plateau—had
the seismograph been discarded—it might have taken the
explorers up to forty years or longer to discover the fields
that geology alone could not find. It was not until war came
in 1939 that the nation, worriedly taking stock of its resources, realized that oil was *the* indispensable material.

What would the outcome of World War II have been if
American oil production had been only half what it was?
Like the first war, World War II was—in Admiral Nimitz'
words—"a matter of oil, bullets and beans."

BOOK FOUR

The
Dynamic
Art

*Prospecting for oil is a dynamic art—
a series of individual techniques—
sometimes overlapping, sometimes sep-
arated by a time gap—but techniques
which lose their usefulness and which
leave us without guide to our prospecting
until we devise some new method. The
greatest single element in all prospecting, past, present, and future—is the
man willing to take a chance.*

E. DE GOLYER

1 : A crooked mile

In the midst of their excitement over the seismograph method, the explorers were given another tool. Unexpectedly, this one was the result of a catastrophe.

On January 8, 1933, workers in the Conroe field, about thirty miles northwest of Houston, were going about their tasks in routine fashion. There really wasn't much to do. The development of this year-old field was well organized. Unlike East Texas, there was no forest of derricks, even though the operators already suspected what soon would prove to be true—that Conroe was a giant of giants.

Without warning, the January calm changed to panic. One of the wells blew out, catching fire. The heroic battle of fire fighting and controlling the well was successfully concluded, and anxious owners of other wells in the field were relaxing, when a second well blew out in a startling fashion. It did not catch fire. The derrick collapsed and the "Christmas tree," an assembly of valves to control oil flow, the name deriving from its shape, disappeared into a great hole in the earth which quickly filled with oil.

Like the crater of a volcano, the hole erupted 10,000 barrels of oil a day. Before the astonished owners could decide what to do, the crater ceased to flow. Four months later it erupted again, and this time it continued flowing.

As Harrison and Abercrombie, owners of the 15-acre lease containing the cratered well, gathered and sold approximately 10,000 barrels of oil a day to keep the lake of oil from spreading, the other owners' alarm turned to dismay. They were legally permitted to produce only 100 barrels a day from each of their wells. The cratered well was not only a

287

dangerous fire threat; it obviously would drain the field so rapidly that it would greatly reduce the amount they would recover under their leases.

The Texas Railroad Commission called an emergency hearing in Houston to discuss the calamity. John R. Suman, vice-president of Humble, which owned almost half the field, made a startling proposal.

"We think we can kill the crater by drilling another well," he said. "We'll drill it four hundred feet away from the crater and slant our hole so that we'll drill into or near the bottom of the other hole. Then we'll pump water or mud down our hole and stop the flow of oil where it starts."

There were incredulous murmurs and lifted eyebrows. The bottom of the cratered hole was 5,100 feet underground. How could an intentionally crooked hole be drilled to come anywhere near such a tiny objective as a little hole only a few inches in diameter?

Texas Railroad Commissioner R. D. Parker swore in as witnesses each of the oilmen at the meeting. Could such a thing be done? he asked. Every witness said no.

John Suman adamantly insisted the scheme would work. Since no one had an alternative, the perplexed Railroad Commission granted permission to make the attempt.

Harrison and Abercrombie agreed to sell Humble the 15 acres containing the cratered well for $300,000, providing they received all the oil the crater produced until the flow stopped, and that Humble would pay for the destroyed oil if the crater caught fire while the relief well was being drilled.

It was an expensive as well as a fantastic gamble, but Humble had a reason for being confident. A young Oklahoman, John Eastman, had drilled intentionally crooked holes two years before at Huntington Beach, California, to produce oil from part of a structure that lay under the ocean bottom. When Suman asked Eastman if he thought he could drill to the bottom of the cratered well, Eastman replied, "If I don't come within fifty feet of bottom hole, you won't owe me a cent." This would be a dramatic opportunity to prove that the technique he had invented was practical.

Elaborate precautions to prevent the crater from catching fire from the drilling well intensified excitement over the unique project. A strong wire fence was built around the 15 acres. Nine guards were continuously on watch. Anyone entering the area was searched for matches. Near the relief well a battery of boilers was erected. Three would furnish steam for drilling; steam from the other six would keep the crater covered at all times with a protective layer of moisture. Three storage houses were stocked with foamite. Safety engineers estimated they could put out a fire in eleven minutes. A corrugated-iron fire shield, 50 feet high and 80 feet long, with a sprinkling system, was set up between the drilling well and the crater. It would protect the crater from sparks, and protect the drilling crew if the crater caught fire.

The plan was to drill a straight hole down to 1,960 feet and then start deviating. Drilling crooked holes unintentionally had been the bane of the industry from its beginning. In the early days in Pennsylvania a well being drilled on a hilltop started producing beer—the hole had sheered out the hill and into the pressure storage tank of a brewery!

In boom fields where rig floors almost touched, the tools often tangled underground in the race to reach oil. If a "straight" hole drilled 6,000 feet deviated as much as three degrees from perpendicular it could bottom anywhere within 7 acres. This could mean that it would produce from somebody else's lease. It could also mean missing the oil formation entirely if the hole was begun on the edge of the pool.

John Eastman had been a salesman, but during the depression he had gone to work in the California fields. The problem of drilling straight holes intriguing him, he invented a device to keep track of the hole's direction and a tool to put it back on a straight course if it strayed. At Huntington Beach, where the field extended under the ocean, he saw wells drilled from the ends of costly piers or on man-made islands. This prompted his revolutionary idea of drilling an intentionally crooked hole from a land site, now that he had tools to make a hole go any direction he wished. His success had gone unnoticed by the industry until now.

When the Conroe relief well reached 1,960 feet, Eastman began deflecting it. When the hole was finished, after two suspenseful months, it was in the oil sand within 22 feet of the bottom of the crater well's hole.

Pumps forced water down the new hole at the rate of 12,000 barrels a day. After two days the boiling surface of the crater became calm, showing that the oil was no longer flowing. Five days later the crater began flowing water, after having flowed 1,500,000 barrels of oil. The vast input of water had pushed the oil back into the producing formation. With the flow of oil stopped and the crater emptied, the damage could be repaired and the hole effectively plugged.

Many a party has been given to celebrate oil flowing from a hole, but only one party has ever been given to celebrate the intentional stopping of a flow. George Strake, the independent Houston oilman who discovered the Conroe field, had told John Suman, "If you kill that well, I'll throw you the biggest party you've ever seen and you can invite anyone you like."

Suman's large guest list was headed by all the oilmen who had testified under oath that the drilling of a directional well to kill the crater was impossible.

No one enjoyed the lavish affair at Houston's River Oaks Country Club more than its host. George Strake had reason to be happy that, instead of being ruined, the Conroe field would produce for many years, proving to contain 747 million barrels of oil, the nation's nineteenth largest field. He had sold his Conroe holdings to Humble for $4 million cash and $3½ million out of a quarter of the oil. Continuing to wildcat, he would eventually pyramid this into one of America's biggest fortunes.

However, his party that night had an additional meaning to George Strake. He was celebrating a satisfaction similar to John Suman's in entertaining skeptics who had been proved wrong.

In 1931, when Strake wanted to drill at Conroe, he could find no one who believed in its possibilities. Gas seeps had been noted in the area as early as 1919, but oddly, these signposts were ignored. Strake couldn't have picked a poorer time to seek wildcatting partners. East Texas had plummeted the price of oil to 10 cents a barrel, and martial law was in effect in both East Texas and Oklahoma. With the dogged determination that is one of the wildcatter's chief assets, George Strake begged and borrowed enough to drill.

In December, 1931, when he completed a well that produced 15 million cubic feet of gas a day, but no oil, major companies and independents continued to ignore the discovery. He managed to drill another well a half-mile south and six months later completed it for 900 barrels of oil a day. Even then, there was little interest, the geologists rating the new field a minor discovery. It took another six months and the completion of one big producer after another before oilmen realized that Conroe was important.

Now, after the sensational crater and John Eastman's incredible drilling feat, there was not an oilman in the industry who was unaware of Conroe. George Strake was only sorry John Suman hadn't invited more people.

Proof that drilling crooked holes could be an art, not an accident, opened exciting new possibilities for the explorers. Natural or man-made barriers to a likely prospect could be circumvented.

Hunting oil on salt dome flanks was the trickiest and most expensive type of exploration because the dome's shape was unpredictable. Now, when a well drilled into salt, the hole could be angled away from it, literally exploring the shape of the dome.

One group, trying to find flank production, penetrated salt at 4,000 feet. Thinking oil might be higher up and closer in to the dome, they plugged the well back to 1,700 feet and started drilling to the side. At 2,800 feet they were in salt again—they were too high on the dome. Next, they angled away from the original hole at 1,900 feet, aiming for a point that would be 80 feet away from the first hole when it reached 3,000 feet. This time, on the third try from 1,700 or more feet down, they found oil.

Ordinary drilling in the Mississippi River where it flowed through Louisiana would have interfered with river traffic. Directionally drilled wells explored the channel from the banks and found new oil fields.

A giant field was discovered at Salem, Illinois, part of it lying under the city's water-supply lake. Directionally drilled wells developed these reserves without contamination. Ceme-

teries were no longer forbidden prospects. Oil could be produced from under Oklahoma's state capitol.

When the Hastings field, another giant, was discovered 15
miles south of Houston a year after the Conroe crater was
quelled, Humble had a lease in the field, but part of it was
a valuable and highly prized fig orchard. Humble drilled
under the orchard rather than in it.

2 : *"Do not buy properties, buy brains"*

W<small>ALLACE</small> <small>PRATT</small> brooded over the maps on his desk. The
evidence was pitifully little, and he had to make an intelli-
gent decision as to whether there might be oil under this
great South Texas cattle ranch. Moments like this depressed
him deeply. When he had to recommend or reject a prospect,
he was never aware of how much he knew, only frustrated
by the incompleteness of his knowledge.

Would he, Pratt wondered, be justified in turning down this
proposal brought to Humble by Hugh Roy Cullen? No
geologist, Cullen was certainly an oil finder. Like so many
wildcatters, he had learned the business the discouraging
way, just managing to keep himself and his family going for
a dozen years while he bought and sold leases and promoted dry holes and small producers. Then he had turned
to exploring abandoned salt domes, and by going deeper had
opened up a series of rich new fields.

Only two years before, in 1932, Humble had paid Roy
Cullen and his partner, Jim West, $20 million for their half-
interest in the Thompson field, which was proving to be

another giant. It was curious how Roy Cullen had drilled
the discovery well in that field. He hadn't even thought of
oil being there. It was Gulf Oil that wanted to lease the land,
but the owner would lease only on the condition that Cullen
drilled the well—and he had given Cullen a lease on half
his land. Cullen deserved the kind of reputation that dumped
a giant oil field in his lap. He inspired confidence and he
certainly seemed to have an affinity for oil.

Wallace Pratt studied his map again. The property itself
was fascinating, oil or no oil. It was the vast Tom O'Connor
ranch, one of the mysterious, unknown cattle kingdoms dat-
ing back a hundred years. The O'Connors, like their neighbors the Kings, ruled their land imperiously, rejecting oilmen
as trespassers.

Pratt could appreciate Roy Cullen's achievement in persuading Tom O'Connor to lease a small part of his domain,
15,000 acres. After fifteen years' work Pratt had just managed to lease the even vaster King Ranch for Humble. I
couldn't have swung it, he thought, if it hadn't been for death
and the depression, and I'm about the only one who believes
there's any oil there.

Why had he persisted for fifteen years in thinking there
was oil under the King Ranch? Was this why he was so reluctant to reject the possibility of oil under the nearby
O'Connor ranch?

In the early twenties there had been an unexciting discovery near the King Ranch, and Humble had managed to
lease a small part of the ranch. A few shallow wells had
found nothing more. Wallace Pratt believed it was good
potential oil land, but only if they could explore the whole
ranch. He had tried to convince three large companies to
join Humble in obtaining a lease, but they all condemned it

as an area of little promise. Furthermore, the King family was suspicious of Humble's intentions and refused to lease the entire ranch.

Mrs. Henrietta King, widow of the river-boat captain who had founded the ranch, died, and the family, faced with paying a huge inheritance tax during the depression, came to Pratt to see how much cash a lease would bring. Again, but without success, Pratt tried to get other companies to join Humble. Even his Humble associates were cool to the idea, but Pratt worked out such an extraordinary deal with the King family that the other Humble executives could not refuse it.

In return for a twenty-year lease, with no drilling obligations, on the ranch's million acres, Humble loaned the King family $3,500,000 at no interest, the loan to come due at the end of the lease. When the family bargained for an additional half million, they received the loan but were required to pay interest on it.

The lease terms were meaningless however, if there was no oil. Even if oil was there, finding it might be a long, costly procedure. The average field covers no more than a 1,000 acres and some are as small as 40 acres. The question of how many dry holes could be drilled in a million acres was mathematically discouraging.

With the King Ranch to explore, why, Pratt asked himself, should he be so perplexed about exploring 15,000 acres on the O'Connor ranch? Comparatively, Roy Cullen's prospect was too small to consider. Cullen thought there was a salt dome there. Wallace Pratt's geological and geophysical staff thought there wasn't, though there might be a structure—or again, there might not. That was the trouble: nobody knew anything.

Cullen proposed that Humble pay for all the geophysical work he had done and drill three wells for a half-interest. It wasn't an enticing proposal, but if Humble didn't accept it someone else undoubtedly would. The wells would be drilled because one man, Roy Cullen, believed oil was there.

One man? No, there were two. For no reason that Wallace Pratt could ever explain to himself, he decided to accept Cullen's proposal. If he hadn't, someone else, not Humble, would have been the other half-owner of the 700-million-barrel Tom O'Connor oil field, the tenth largest in Texas.

Nor was Pratt's faith in the King Ranch unrewarded. While Humble was seismographing the ranch, wildcats drilled outside the property discovered several fields, all of which extended across the lines. Humble's first King wildcat discovered a major field. In the years since, the extent of the cattle empire's underground riches has still not been completely determined.

It was his faculty of never letting what he knew get in the way of what he didn't know that made Wallace Pratt an oil finder. As a scientist he shared with the practical wildcatter the willingness to take a chance. This is why Humble Oil and Refining owned either all or a great part of eight of the twelve giant fields discovered in Texas between 1930 and 1935. Four of them were Humble discoveries and four of them Humble bought after their discovery by independents, recognizing their potentialities so quickly that, in effect, they bought hundreds of millions of barrels for only a few cents a barrel. During this period Texas and the nation doubled their discovered oil reserves, but Humble increased its reserves thirteenfold, double those of its nearest competitor.

"Do not buy properties, buy brains," was John D. Rockefeller's motto as he gathered in companies to form his Standard Oil Trust. If Jersey Standard president Walter Teagle had this in mind when he bought a half-interest in the fledgling Humble company in 1919, he had cause to doubt his judgment after the Ranger debacle of 1921 when his associates referred to Humble as "Teagle's sick baby." But by 1933 the baby was dominating the parent. Humble brains effected the final transformation of Jersey Standard from primarily a refining and marketing company to one in which exploration and production played a major role.

Companies change direction according to the men at the top. In a big company it is never a sharp change. Just as a great river like the Mississippi describes a huge oxbow in starting to change course, so a big company takes a broad, slow swing.

In 1933, Humble president Bill Farish, the independent-oilman product of Spindletop, became chairman of the board of Jersey Standard. Four years later he switched places with Walter Teagle, becoming president.

Bill Farish brought Wallace Pratt to New York as a Jersey Standard director and member of the executive committee. The emergence of a geologist as one of the guiding minds of company policy would have shocked the original Standard Oil board members, particularly the one who always stated at every meeting, "Remember, gentlemen, we are only merchants."

Being a merchant was no longer enough. A new generation of scientist-businessmen was rising to the top. In the early twenties Wallace Pratt hired two young geologists for Humble, fostering them as Daddy Haworth had once fostered him. Eugene Holman and Leonard F. McCollum were both

University of Texas graduates. They absorbed Pratt's philosophy of the art of prospecting, but unlike him, they both had a talent and liking for administrative work.

Humble was expanding so rapidly and there was so much to be done that the company was fertile ground for their growth. Responsibility was practically thrown at them. When Bill Farish assigned Gene Holman the job of superintending Humble's Louisiana-Arkansas division, his only instructions were, "Gene, go up there and make mistakes, but for Christ's sake, don't make too many of them."

Gene Holman didn't. In a few years he was in New York advising Jersey Standard on exploration problems and organizing its exploration company in Venezuela, which nation became Jersey Standard's greatest source of oil production.

John D. Rockefeller would have been startled when, in 1942, Gene Holman, at forty-nine, became president of Jersey Standard. In 1954 he became chairman of the board. Charles Gould's key question of 1912—"Why bother with a geologist at all?"—had received its most dramatic answer.

By 1959 more than four-fifths of the fifty largest oil companies in America had a geologist as either president or vice-president.

One of these presidents, giving Jersey Standard some of its keenest competition, was Wallace Pratt's other protégé, Leonard McCollum, president of Continental Oil Company. "Mc" McCollum was transferred from Humble to Carter Oil Company, Jersey Standard's mid-continent subsidiary. First as exploration manager and then as president he became noted for being "fustest with the mostest" with his new leasing technique.

After deciding to explore a new area he would plan a secret leasing campaign and strike with lightning speed. He would assign as many as a hundred men to the operation, organized in combat teams of three to lease specified areas. Team captains would report progress to a crew chief. Crew chiefs reported by telephone to headquarters in Tulsa where, in a closely guarded conference room, leases acquired were marked on a huge map so that Mc McCollum and other executives could follow the play.

In the field, the lease men took leases in their own names and paid bonuses with cashier's checks to conceal their company's identity from competitors. They worked so rapidly that Carter would have the majority of the land it wanted before other companies realized what was happening.

Jersey Standard brought McCollum to New York as vice-president in charge of its world-wide production. Wallace Pratt and his protégés were all under one roof again, but not for long.

Continental Oil's hard-boiled policies in reorganizing E. W. Marland's life work had resulted in hardening of the arteries. The company's heart was gone. Given the presidency, Mc McCollum in 1947 began the work of bringing it to life. Had Marland lived to see McCollum's achievements, his bitterness would have been softened by admiration. McCollum figuratively and literally tore down a wall between executives and employees. Marland, with patriarchal lavishness had built a magnificent recreation building with dining rooms, cafeteria, gymnasium, theater, and lounges across the street from the company office building. To join the two, he had built an enclosed bridge over the street. When Dan Moran took charge of the company, after Marland's ouster, he ordered the bridgeway partitioned so that executives would not have to mingle with employees.

McCollum's first executive order called for tearing down

the partition. The day after it was removed he was walking across to lunch when an office boy fell into step with him. "Isn't it wonderful?" the office boy commented. "God is dead."

At that moment someone said, "Hello, Mr. McCollum." With one startled look the office boy turned and fled.

With the same generalship he had shown in his leasing campaigns, McCollum increased Continental's gross annual income from $228 million to more than $3 billion at the time of his retirement in 1970. The mark of the prospector and geologist could be seen in the company's lease figures of undeveloped petroleum acreage that year. Beginning with 2 million acres, McCollum raised Continental's total to more than 300 million acres.

3 : *"Prospecting is like gin rummy"*

IT TAKES LUCK to find oil," Everette DeGolyer once remarked. "Prospecting is like gin rummy. Luck enough will win but not skill alone. Best of all are luck and skill in proper proportion, but don't ask what the proportion should be. In case of doubt, weigh mine with luck."

A Houston wildcatter, Jack W. Frazier, would have agreed that the case could not be stated better. He had been drilling one dry hole after another for sixteen years when, in 1935, his luck seemed to change. During the next five years he discovered almost a dozen fields—not big ones, but big enough to make his participation in them worth several million dollars. Now, he thought, there would be no stopping him.

Going ahead with enthusiasm and assurance, Frazier drilled prospects recommended by geologists or seismograph work. Each one had the possibility of discovering another giant field, but none of them did. From January 1, 1940, to January 1, 1950, Frazier drilled 200 wildcats and lost $2 million on 188 dry holes. The heartbreaking thing was that he lost a quarter of a million dollars on his twelve discoveries. They were good enough to warrant completing the wells, building storage, and beginning production. The only difficulty was they didn't produce enough to repay the drilling costs. Frazier experienced so much trouble trying to produce one of the wells that after a year, when he gave up, the oil he had produced had cost him $1,000 a barrel.

He didn't quit. He continued to pour back into the ground the money coming from his earlier discoveries. Later, he hit again, not a major field, but enough to give him new capital to continue his search for that giant oil field, the thought of which keeps a wildcatter going.

Jack Frazier knew the odds against his finding such a field. His "stinkers," as he called the ones that found him a little unprofitable production, were among those listed in the statistics as the one out of 9 wildcats that find oil. Since the 1940's, to discover a million-barrel field a man may have to drill 50 wildcats. At this point his chances become slim. Only one wildcat in 750 discovers a field of 10 million or more barrels, only one in 2000 discovers a field of 25 million barrels, and only one in 3000 discovers a field of 50 million or more. As for the giants, containing more than 100 million barrels, less than 300 have been found since America's first oil well.

What keeps Jack Frazier, and thousands of less successful operators, wildcatting in the face of such fantastic odds?

Simply the knowledge that the rewards for success are commensurate with the risk. Oil is the greatest single source of wealth in America for individual fortunes. At the same time, exploring for it is the greatest source of business failure, a fact to which wildcatters deliberately blind themselves. They disregard the unfulfilled dreams and broken lives that lie buried at the bottom of the staggering total of 730,000 dry holes drilled in America, and think only of those who, despite every difficulty, persevered to success.

Of all the wildcatters who won, Haralson L. Hunt's success was weighted more strongly with luck than skill.

A professional gambler seldom makes a risky bet, but after Hunt parlayed the Arkansas lease he won in a poker game to the purchase of Dad Joiner's holdings in East Texas, he became the wildest wildcatter of them all. His luck was so astounding that by 1950 he was considered the richest man in America.

H. L. Hunt was contemptuous of geology and geologists. He refused to look at their maps. "Finding oil is all luck," he told a friend. "Some people are born lucky and are lucky all their lives. Others are born unlucky."

There seemed to be no room for argument in his case. If oil companies were leasing prospective land in one direction, Hunt would lease in the opposite direction, for no reason except that the leases were cheaper because there was no competition. Oil would usually be found on Hunt's leases.

He would not drill on scientific recommendation, but might have forty wildcats drilling at one time on hunches or even a dream.

Atlantic Oil leased and seismographed 2,000 acres in Louisiana. Deciding the prospect was not important enough, the company tried to get someone to drill, keeping a royalty. Nobody else liked it either, including H. L. Hunt's company.

Jimmy Owens, a drilling contractor, went to see Hunt personally. "Mr. Hunt, I had a dream last night," he said. "There's a play over in Louisiana I know about and I dreamed I drilled a well on it for you and we hit. You and I were out at the well sitting on a log playing poker and I won $500,000 from you. I took the check to the bank to deposit and my banker asked me what I was going to do with the money. I told him I was going to buy another rig and go to Canada and he said I had enough pig iron already and I'd better pay off my notes."

Hunt's green eyes sparkled. "Well, if you won that much money from me in a poker game, Jimmy, there must be oil there." He laughed. "We'll just drill it and see. I'll carry you for a half-interest and give you the drilling contract and when we hit oil you can drill up the field."

It was no surprise to H. L. Hunt when the well found oil, not just a little but one of those 750 to one shots—a 15-million-barrel field.

Hunt ran his company out of his hip pocket. No man knew what another was doing. Hunt was a hunter whose pack of bird dogs worked in circles. Their purpose was to flush the birds for him. One day in the late forties, he showed up unexpectedly in an office the company had in Midland, Texas.

A rising young oilman, Jimmy McRae, was unsuccessfully trying to sell Hunt's land man a drilling deal. Hunt nodded amiably and remarked quietly to the land man, "I want you to spend four million dollars drilling wells during the next sixty days."

The land man gasped. "But, Mr. Hunt, we don't have any

place to spend it and we couldn't possibly spend that much money in that short a time."

"Just spend it drilling," Hunt said affably, and walked out.

The land man turned despairingly to Jimmy McRae. "Well, Jimmy, I'll take that deal. It didn't look good a few minutes ago, but it sure does now."

The land man telephoned a lease broker in Houston. "Pass the word to all the boys to round up their drilling deals and meet me at the Rice Hotel in two days. I'll take four million dollars worth of deals."

Jimmy McRae grabbed his hat. "I'll see you at the Rice, myself," he said.

At the hotel, the promoters of oil deals stood in line. Two secretaries and a lawyer flanked the land man. Disgustedly, he signed contracts for every proposition, no matter what, as long as the lease titles were good and the wells could be drilled immediately. Every cat-and-dog deal that had been floating around unaccepted for years was about to have its chance.

The campaign was the joke of oil circles. Everybody knew that Hunt was in a hurry to spend money before the end of the year, since, if not used for drilling, the money would have to be paid in taxes. "What a stupid way to throw away money," one oilman said. "If I just had that money to spend on some really good prospects, I'd find oil."

H. L. Hunt didn't mind being called stupid. When his several hundred wildcats were drilled, he almost doubled his own oil reserves, and the nation was enriched with new reserves that otherwise might not have been found.

Some of H. L. Hunt's "luck" rubbed off on Jimmy McRae. On a smaller scale he, too, parlayed seemingly unpromising prospects into a successful, aggressive independent oil exploration company. He disagreed, however, with Hunt's belief that it was all luck. "What Mr. Hunt proves and what every independent oil company proves is that if you're willing to take a chance and drill enough wells without going broke, you're bound to hit," he said. "It's drilling the wells that counts. The more money you spend drilling, the more certain you are to find oil in spite of the odds. Once you're on top it's not as much of a gamble as it seems. But when you're on the bottom or in the middle you're betting your shirt and there's only one H. L. Hunt for every thousand wildcatters who don't make it to the top."

Hunt gambled on everything. On football weekends he would bet $25,000 a game on each of six games. He kept two accountants busy figuring odds on horses so he could bet on practically every race in the country. One year he won a million dollars on racing bets.

With the exception of Mike Benedum, all the great explorers have avidly sought some form of gambling as recreation. In explaining E. W. Marland's passion for big-stake poker, his biographer, John Joseph Mathews, writes in *Life and Death of an Oilman:* "Every species of animal that defends itself, or depends upon the overtaking or destruction of others for food, must spend much time playing. The play of such animals is utilitarian, since in it they re-enact the manner in which they defend themselves or the craft by which they take their prey. A cougar plays with a saddle blanket lost by a pack outfit in the mountains, and the house cat attacks a dead mouse or a ball of yarn. Rabbits and squirrels, the hunted, play at chasing each other and taking refuge. Acquisitive men usually play those games wherein they seem to be keeping bright their power to remain in the struggle."

Harry Sinclair, as inveterate a recreational gambler as H. L. Hunt, once demonstrated the value of keeping his power bright in a very literal sense. In his early wildcatting days he was drilling a well and did not have the money to meet the Saturday payroll. He dropped by the rig and casually suggested to the crew that they have a little dice game. In an hour or so he won enough money from them to pay their wages.

Mike Benedum considered recreational gambling a non-constructive use of money. For him, the only worth-while gamble was a hole in the earth that would enrich everyone, not just himself, if he won. Nor did Benedum think that luck played any special part in his oil successes. As an associate remarked: "I watched Mike Benedum's luck. It consisted of having his feet under the desk every morning around 9:30 and being there until 4:30 every afternoon. It consisted in his knowing every well that he was drilling, taking an interest in his companies, plus the exertion of considerable ability in the way of a great brain and a great personality."

That two such different personalities as Mike Benedum and H. L. Hunt, using entirely different approaches, should each be a great oil finder, demonstrates that in oil finding the basic ingredients of "luck" are the willingness to take a chance, persistence, and the ability to think and act in unorthodox fashion. These qualities explain the fact that independent oilmen have consistently discovered three-fourths of all new oil fields.

The great geologists have been successful only in the degree to which they, too, have been unorthodox in their thinking. In the early 1930's DeGolyer demonstrated this with seismograph, and Wallace Pratt's unorthodox thinking made Humble a great producing company. In Oklahoma, another geologist, Arville Irving Levorsen, using the same principle, would also become one of the great shapers of exploration thinking throughout the world.

In 1931, however, philosophical principles were decidedly secondary to paying the grocery bill as far as Levorsen was concerned. At thirty-seven, he was broke, out of a job, and worried about how he would care for his wife and four children.

A tall, gentle, humorous Norwegian, Levorsen had been a geologist for several independent oil companies since his graduation from the University of Minnesota School of Mines. He had not been especially concerned when the company he was working for sold its properties in 1930. Then he learned that the depression and the flood of East Texas and Oklahoma City oil had left the companies in no mood to hire geologists.

Early in his career, Levorsen reduced the principle of exploring for oil and gas to a profound simplicity, which he stated in six words: "Find a trap—then drill it."

He thought about this as he sat in the Tulsa stock brokerage office where he and other geologists and independent oilmen who could not afford an office often assembled to keep warm, look at each other's maps, and exchange ideas. The only charge for this haven was an occasional interested look at the ticker tape. If finding oil was as simple as he had always maintained it was, Levorsen told himself, now was the time to prove it.

"Find a trap—then drill it," he mused. "Perhaps I'd better concentrate on the first three words."

The boys in the brokerage office saw him less frequently. Levorsen was spending his time in the library studying geological and production maps. He was searching oil company files, examining logs of hundreds of wells. He sat in his study late every evening poring over maps and data.

"I'm looking for an idea," he explained to his wife. "Any undiscovered oil or gas field at best exists only as an idea in the mind of the geologist. Imagination is what it takes. Thousands of wells have been drilled in Oklahoma, but look at the blank spots on the map. I'm convinced that I'll be able to see an oil field in one of these blank spots, if I correctly interpret the story told by the wells which have been drilled. Somewhere I'll find a clue."

At two o'clock one morning, Levorsen could no longer see one of the blank spots on the map. It had been replaced in his mind by an oil field. The production nearest it was 32 miles away. Eagerly he reviewed all the facts he had been juggling. No doubt about it—the oil field was there. Now the problem was: how would he manage to lease and drill?

He was again a frequent visitor in the brokerage office, but after raising the lease money, he seemed no nearer to drilling than before. Every oil company turned the prospect down. They could not see the field in Levorsen's mind. The very facts that made him believe in its existence were to others equally valid proof that oil could not be there.

Levorsen presented his case to Ed Moore, a lawyer he knew. Moore had inadvertently found himself in the oil business by being forced to accept leases and interests from clients who couldn't pay their bills. His partner, a grocery-store owner, was in the business for the same reason.

Moore was not handicapped by thinking he knew how to find oil. If Levorsen saw an oil field, it must be there. Agreeing to drill, he paid back the money already spent on the leases. Levorsen received a thousand dollars and a participation. The thousand dollars looked bigger than the oil field.

Moore did not have enough money to drill, but he had caught enough of Levorsen's vision so that just before the leases expired he mortgaged his home. They found only a small amount of oil. Levorsen felt that if they had continued on down to the Wilcox sand, Oklahoma's most prolific producing formation, they would have found his pool.

Still believing, Ed Moore borrowed more money to drill a second well. When the bit penetrated the Wilcox sand, another giant oil field was brought to life, the Fitts pool, containing 190 million barrels of oil.

The discovery gave Levorsen economic freedom to do what he was superbly equipped to do—stimulate the creative thinking of all geologists through research, lecturing, and teaching. He became Dean of the Mineral Sciences at Stanford University, and he wrote the outstanding textbook on petroleum geology. As a consulting geologist for foreign governments and oil companies, he generated new ideas for new oil provinces—and he continued to preach the gospel of unorthodox thinking as the scientist's most valuable oil-finding tool.

Only infrequently in a major company is it possible for an unorthodox idea or the playing of a hunch to rise above the hampering restrictions of group thinking. When it does, it may result in brilliant discoveries.

H. S. M. Burns, a Scot whose personality and wit are second only to his ability as a geologist and geophysicist, became president of Shell Oil Company in America in 1947. He had headed Shell Oil Company of Colombia but most of his experience had been with Shell of California.

The first day he took over in the New York office, he found a pile of papers on his desk awaiting his signature. On top was an order, approved by all the appropriate committees, to abandon a block of leases in western Oklahoma.

Max Burns read the report. The block had been seismographed and a test had been drilled to 13,000 feet. It was a dry hole. No big oil fields had been discovered in this part of Oklahoma. The sensible thing was to forget a failure. Burns was expected to approve the committees' judgment.

The explorer in Max Burns revolted. "I don't know why, but I couldn't begin my job as president by condemning an area as having no oil," he said. He substituted an order to re-examine the prospect. His hunch, his unorthodox procedure, hit the rare jackpot—a giant field. A shallower test demonstrated that the first one had drilled right through the producing sand. The drilling mud had sealed off the flow of oil. The leases Max Burns impulsively refused to abandon covered almost all of the 120-million barrel Elk City field.

The creative geologist who thinks in terms of great oil provinces rather than a specific small area has an unusually difficult time getting group thinking to translate ideas into action.

Lewis Weeks, a Wisconsin geologist, stretched the horizons of his mind by studying world petroleum geology. He had been in Mexico and India before he joined Jersey Standard in 1924. After fourteen years in South America, working in Venezuela, Colombia, and then as chief geologist of Standard Oil of Argentina and president of Standard Oil of Bolivia, he became Jersey Standard's senior research geologist in New York.

Lewis Weeks was in search of great untested regions where the geological environment would promise the existence of many oil pools. The habitat of oil throughout the world is in basins where large volumes of sediments have collected with favorable conditions for the accumulation of oil. Weeks's attention focused on two closely related basins— the Alberta, in Western Canada, and, to the south of it, the Williston, in North Dakota, South Dakota, and Montana. Geologists knew the basins but no one had suggested they were great oil provinces. Although some oil had been found, results in general had not indicated the basins were good hunting grounds.

Weeks studied the vast area, looking for oil-source rocks and those that were porous and permeable enough to serve as a reservoir, looking for clues to geological events that would have formed traps underground. In Canada, on the banks of the Athabaska River, he examined the thick tar sands that oozed a black, viscous oil. It seemed incredible no method had yet been devised to process economically this largest known single deposit of oil in the world, containing 300 billion barrels. This was the residue of a fabulous pool. Where there had been such vast amounts of oil, there still must be other accumulations in older rocks of Western Canada.

Painstakingly, carefully, Weeks cemented bits and pieces of evidence into theories and arguments, illuminated with imaginative, creative thinking, as to why the Alberta and Williston basins had vast oil potentialities. He discussed his ideas with Wallace Pratt and other Jersey Standard geologists, converting them to his belief.

Like Pratt, Weeks was a philosopher and an independent spirit. He planted himself on his convictions, but for six frustrating years Jersey Standard management would not venture with the drill into these unexplored provinces, feeling it was too great an economic risk.

In 1946, Jersey Standard's Canadian affiliate, Imperial Oil Company, held a symposium of experts—pipeliners, econ-

omists, marketers, exploration men—to determine where
Imperial could develop a billion barrels of oil that would
be accessible and transportable.

Mose Knebel, the Jersey Standard geologist who had seen
the "red flags" in East Texas before Dad Joiner's well, agreed
with Weeks's ideas. Suddenly and dramatically he caught the
group's attention by slapping his hands on the map of
Canada.

"Lewis, you've been talking so much about the possibilities of the Alberta basin. Where would you select acreage?"

Weeks quickly outlined a long rectangle extending north
and south through Alberta. The pipeliners, economists, and
marketers could see no objections.

A leasing and seismograph campaign started. A year later
the first wildcat discovered the Leduc field, setting off one
of the world's great oil booms in Western Canada. Lewis
Weeks had attained the geologist's ultimate goal: the discovery of a pool that opens up a new petroleum province.

The discovery was not the bonanza for Jersey Standard it
could have been. Instead of boldly leasing all the area Weeks
outlined, the company leased only a fraction, leaving a whole
series of lucrative pools for others to discover.

Even though Weeks's theory concerning the Alberta basin
was proved, management did not act on his recommendations of the Williston basin to the south. In 1951 Amerada
drilled the first discovery there, opening another oil province
of 30 million acres.

The reluctance of the major companies to explore frontier
areas in the 1950's was based on the knowledge that half of all
money spent on exploration and drilling resulted in no produc-

tion. It was less risky to buy acreage in a new area after a discovery had proved it was oil territory. The least risk was to buy the discovery.

As a whole, geologists became as conservative as the economists. Knowing the inexactness of their science, they showed increasing reluctance to recommend drilling an imaginative wildcat. After the easier, shallow fields were discovered, wells were drilled deeper and deeper until the average 5,000-foot wildcat costs $200,000 and some even a million dollars. As the search for oil became more expensive, more complicated, a simple fact became clear. Only the rare geologist has the creative mind that can take hold of obscure and miscellaneous data and by deduction and intuition arrive at a new idea of where oil may be. One executive estimates that only one out of twenty geologists is an oil finder.

Few of today's geologists clear their thinking with the cheerful, candid approach of Glen Ruby. "Finding oil is simply reading tea leaves through a crystal ball," he said. "If you want to find oil throw away the textbooks and go out in the field."

Following his own advice, Glen Ruby was a geologist who was an oil finder. He made the discoveries farthest north and farthest south in the Western Hemisphere: in Alaska for the United States Navy, in Chile for the Chilean government.

"If a man can't find oil for himself he shouldn't be going around telling other people where to find it," Ruby announced at the height of a career that included his being consultant for more foreign governments than any other geologist. Before becoming an independent consultant, Ruby was E. W. Marland's exploration manager of the Hudson's Bay Company lands in Canada, resigning when the bankers took over.

During all the years he roved the world's petroleum provinces, Ruby could not forget a rugged, arid region in Utah, the northern part of the sprawling Paradox basin. He had tramped over the area in 1934, during his early days with Marland. Even though no oil had been discovered there, major companies and independents were exploring the southern part of the basin. One company, over a 10-year period, drilled fifteen consecutive dry holes and spent 8 million dollars without finding oil.

Glen Ruby was convinced, nevertheless, that the Paradox basin was a great oil province. One anticline, called the Big Flat, haunted him. He leased it and persuaded Tidewater Oil to drill in return for an interest. There was a show of oil, but the engineers were unable to complete the well as a producer. Tidewater abandoned the prospect.

Meantime, exploration efforts in the southern part of the basin succeeded. In 1956, a Texas Company wildcat discovered the Aneth pool, which quickly proved to be a giant. Glen Ruby's conviction that the Paradox basin was a great new oil province was proved, but by someone else and several hundred miles away.

Other geologists thought Ruby was throwing away his money, not his textbook, when he drilled a second time on Big Flat, almost twinning the Tidewater failure. The second failure cost Ruby and some friends $400,000. Again there was a show, but the well could not be made to produce.

Ruby refused to disbelieve his interpretation that originally convinced him Big Flat contained flowing oil in big quantities. Stubbornly determined, he found an independent, King Oil, that was willing to drill yet another well. It was the same story again. Finding another company that would overlook three failures was not easy, but in August, 1957, the fourth try, drilled by Pure Oil, began flowing 500 barrels a day, starting a rush to the northern part of the basin. Glen Ruby had correctly read the tea leaves through the crystal ball and opened a major pool.

Incredible as it seems, the natural resource upon which the world—and particularly the United States—is most dependent, next to food, still cannot be found directly. As we have seen, it must still be found by a combination of random drilling, luck, unorthodox thinking, and the use of techniques that do not have guaranteed results.

Ever since Edwin Drake's discovery at Titusville a hundred years ago, there have been fears of a shortage. Each time the cry of alarm was raised, the explorers' reply was a new wave of discoveries. More oil was found because more people went looking for it. By 1959, the independent exploring and producing companies numbered twenty thousand.

While the independents have discovered four-fifths of all America's oil, they could no more exist without the major companies than the majors could exist without the independents. Half the oil produced by the independents has been sold to the majors, enabling the independents to continue their searches. Small operators rely on big companies to buy the oil they produce, to build the vast network of pipelines for its transportation, to build and operate the costly and highly specialized refineries and petrochemical plants that produce the myriad products that can be made from oil, and finally to market these products on the tremendous scale the nation requires.

"An 'independent' is a misnomer," Wirt Franklin, discoverer of the Healdton field, said at the time he organized the Independent Petroleum Association of America. "There is no class engaged in productive enterprise more dependent than the small oil producer."

Independents are also dependent upon major companies for a great many of their successful prospects. The majors lease large quantities of land in a new area. Retaining the best prospects—determined by preliminary geological and geophysical work—they farm the remainder out to independents willing to drill for a half-interest. Nothing better illustrates the inability of geologists and geophysicists to interpret their information accurately than the fact that so many of these farm-outs discover major fields.

John W. Mecom, a Houston wildcatter, was able to discover more than a hundred million barrels of oil on such farm-outs. After roughnecking in the fields for several years during the depression, he borrowed $700 from his mother and an old wooden derrick from his father, a moderately successful part-time drilling contractor. With two partners, Mecom worked over abandoned wells, getting them to produce again. By 1945, he had a small amount of production. Pan American Petroleum, the exploration and producing subsidiary of Standard of Indiana, owned some acreage near a producing salt dome on the Texas Gulf Coast, but no one believed it would be productive. Mecom, gambling that it would be, took a farm-out and his deep wildcat discovered a new field on the High Island salt dome's flank, putting it into the giant field class.

To the discomfiture of Pan American's exploration department, Mecom subsequently took a farm-out of some of their Louisiana offshore acreage. Buying their seagoing drilling platform for its salvage value, he developed production of more than a thousand barrels of oil a day and 100 billion cubic feet of gas reserves, selling the latter for 16 cents a thousand cubic feet.

Although independents make the majority of discoveries, the major companies, by perfecting production techniques and practices, have been primarily responsible for more than doubling the amounts of oil the discoveries yield. The early fields were produced so wastefully that in some cases only a fifth of the oil contained in the rocks was recovered. The great majority of the reservoir remained unobtainable, inert in the rock pores, after its moving energy, gas, was permitted to escape uncontrolled.

The major companies' research and production engineering departments have devised techniques of injecting water, gas, air, steam, and chemicals into oil-laden rocks to move the remaining oil to the well holes where it can be pumped. They have perfected processes of fracturing and acidizing rock formations where the oil is held so tightly that otherwise it could not be produced.

The major companies became big because of their concentration on those branches of the industry which, unlike exploration, have only normal business risks involved. However, with the exception of six which are offshoots of the old Standard Oil Trust, the twenty largest integrated oil companies were all started by independent explorers and producers, and became majors by competing with Standard companies in transportation, marketing, and refining.

In the group of ten American oil companies with assets of more than a billion dollars, only four are Standard companies, despite their long head start in supremacy of every industry phase except exploration.

The offspring of Spindletop, Gulf Oil and The Texas Com-

pany, equal in value any of the Standard companies except
Jersey Standard. Harry Sinclair's dream of surpassing John D.
Rockefeller made Sinclair Oil the seventh richest and most
powerful company before it was finally merged with Atlantic
Richfield. The group of independent Dutch and British pro-
ducers and refiners who formed Royal Dutch-Shell to fight the
Standard trust, so successfully invaded Standard's home terri-
tory that Shell Oil Company in America is among the top ten.

Cities Service earned its membership in the billion-dollar
club as the result of one of the most agonizing decisions an
executive ever had to make. The diversified genius of Henry
L. Doherty brought together under one roof some 250 utility
corporations as well as his oil and exploration companies that
discovered the giant fields of El Dorado, Seminole, and Oklahoma City. The Public Utility Holding Company Act of
1935 required public utility companies to confine themselves
to a single integrated utility operation in a limited area,
which meant that if Cities Service remained in the utility
business it must dispose of its oil and gas interests.

W. Alton Jones, who started as an auditor in a Cities Service Missouri utility company, rapidly became Doherty's most
valued assistant; at forty-four he was president of Cities
Service. Doherty's death left the painful decision to him.
He chose to sell the utility companies and keep the oil company—and showed as great a genius for expanding the dis-
membered company in all phases of the oil industry as
Doherty had in its creation.

One of the youngest members of the billion-dollar club,
Phillips Petroleum, joined in 1953. Its roots were deep in
the early Oklahoma discoveries, and, like the Sinclair company, it expanded rapidly into all phases of industry operation. Responsible for this development was Frank Phillips,

an Iowa barber and bond salesman, who went to Bartlesville, Oklahoma, in 1904 and started a small bank. Ambitious and aggressive, Phillips bought and sold oil leases on the side in the early booms. Banking soon became the sideline. Organizing an oil company in 1917, Phillips followed the explorers into the Osage and subsequently into the Oklahoma City field. Thereafter he branched into refining and marketing.

In 1945, when Frank Phillips turned over management of the company to K. S. "Boots" Adams, it was already a major one with assets of $317 million. Boots Adams, who had started with the company as a warehouse clerk, tripled its assets in less than ten years, principally by becoming one of the largest manufacturers of petrochemicals and emerging as the nation's number one producer of gas. In the early thirties, gas discoveries were of little or no value, but Adams persuaded Frank Phillips to buy a vast amount of cheap gas acreage in the Panhandle and Hugoton fields. Adams was no longer accused of "chasing gas bubbles" by other company executives when the demand for natural gas doubled after World War II, and gas began its rapid rise as one of the nation's most important sources of energy.

Phillips and the other big integrated companies that rose by competing with Standard bear out Wallace Pratt's observation that "the entire American oil industry is but the lengthened shadow of the independent oil man, whose form and substance are stamped indelibly over its whole structure."

4 : "The profit of the earth is for all"
(Ecclesiastes)

WHEN THE industry celebrated its centennial in 1959, the explorers had found 95 billion barrels of oil and 395 trillion cubic feet of natural gas in America. Ninety-eight per cent of this vast quantity was found after Spindletop's discovery in 1901.

Never before in history had a natural resource enriched a whole nation on such a great scale.

Oil and gas provided two-thirds of the nation's energy requirements. Translated into the factories it ran, the homes it heated, the cooking stoves it fueled, the buses, trucks, cars, planes, and ships it powered, and the jobs it directly and indirectly created, this figure meant that the nation was primarily dependent upon oil and gas for its high standard of living.

Oil liberated a whole class of workers. At the turn of the century, three out of every ten workers were engaged in farming. By 1959, thanks to gasoline-powered tractors, cars, and trucks, only one in ten was a farmer, while the United States was one of the few areas in the world with surplus food. Farmers harvested a $300 million crop annually from the oil companies—the rentals and bonuses for oil leases on prospective land. Only 20 million acres produced oil, but the explorers had 384 million nonproducing acres under lease, more than a fourth of the United States' land area. The owners of the 20 million productive acres—including, as well as farmers and ranchers, schools, churches, federal and state governments, and Indian tribes—received as royalty an average of an eighth of all the oil produced.

When the rich Huntington Beach field was discovered in California, oilmen had to go to Maine, New Hampshire, and Vermont in search of the owners of a great part of it. Many years before, an enterprising encyclopedia salesman had bought the land for almost nothing from a disappointed real-estate promoter and increased the sales of encyclopedias by offering New England farmers a free lot in sunny Southern California with every set.

A desolate tract in West Texas was sold in 10-acre parcels to hundreds of dwellers in New York, Newark, Chicago, and Pittsburgh who didn't realize the palm trees on the brochures grew only in the artist's imagination. A few years later an oil company's land man painstakingly followed up the salesman's visits to obtain leases—the supposedly worthless land was in one of West Texas' lush oil fields.

The reaction of landowners to unexpected riches has been as diverse as the ownership. An unexpectedly large number of farmers refused to let wealth change their way of living.

The extent to which one Illinois farmer indulged himself was to buy a pickup truck and take his cow riding every evening because she seemed to enjoy it so much. His neighbors, an old bachelor and two spinster sisters, had always wanted to go to Florida. Their oil royalty dollars piled up in the bank, but the trip never got beyond the conversation stage: they owned a pair of mules worth $60 and didn't know who would look after them properly while they were gone.

In the middle of the Fitts field in Oklahoma, on 10 neatly fenced acres, was a little cabin where an old farmer lived by himself. He refused bonuses of $150,000 for a lease on the acreage, because it meant his cabin would have to be moved. He had already leased the rest of his farm and more money would have meant nothing to him. He didn't know what to do with what he had. A land man took a picture of the shabby little cabin and sent it as a Christmas card to all his friends who had also tried to get the lease. Under the photograph was printed the question, "How would you like to live in this $150,000 home?"

Most successful oil explorers have been unostentatious with their personal riches. H. L. Hunt and Sid Richardson, the only two explorers whose fortunes were estimated to be among the fifteen largest in America (in the 200 million to 700 million dollar bracket), lived so quietly and modestly that they were never seen except by a small circle of friends.

Hunt built himself a replica of Mount Vernon on the shores of a Dallas lake; his principal recreation was rocking on the porch and musing over his gambling bets. He had a passion for anonymity and rarely traveled, but was lured to the Persian Gulf in 1957 in an effort to outbid a rival, J. Paul Getty, for some valuable concessions in Saudi Arabia. Hunt was not so much interested in the concessions as he was in proving that he was a better oilman than Getty, who had received great publicity as having surpassed Hunt as America's richest oilman, having become the fourth American to be classified as a billionaire, joining the company of Rockefeller, Mellon, and Ford.

Neither Hunt nor Getty won the concessions, as the Japanese outbid them both, nor was there really any doubt as to which of the two was the best oilman. Getty was richer, but his wealth was achieved by acquiring oil companies through stock manipulations. Hunt won his fortune the more difficult way, by exploring for oil.

Sid Richardson, who trailed Hunt in riches, owed the greater part of his fortune to oil. A bachelor, he lived simply in two rooms at the Fort Worth Club, vacationing in Southern

California and on an island he owned in the Gulf of Mexico. His career and fortune were inextricably part of those of his boyhood friend, Clint Murchison, whose estimated $350 million empire was less than Richardson's.

Richardson and Murchison were born in 1891 and 1895 in Athens, a hamlet some 50 miles west of the East Texas field. After World War I they were attracted to the West Texas oil booms. Trading leases and promoting wells, the partners developed production that financed them into the East Texas field, which in turn launched them on a round of wildcatting and rich discoveries. Their primary pleasure was trading, which was why they used their oil success as a springboard to an amazing variety of investments: chemical plants, candy companies, cattle ranches, drugstores, publishing companies, silver factories, railroads, taxi and bus companies, steamship lines, restaurants, newspapers, washing machines, insurance, motels, water companies, banks—anything, in short, that offered the opportunity for a deal and a profit.

Few of the richest oilmen resemble the caricature of the Texas that has entered American folklore: the free-spending, hard-drinking, brawling, Stetson-wearing character with a fleet of Cadillacs—the fellow who, when his dentist examined his teeth and said he could find no cavities, replied heartily: "Go ahead and drill anyway—I feel lucky today!" This concept was kept alive by the antics of a minority—men like Glen McCarthy, a wildcatter who sensationally made and lost several fortunes, but never lost a headline. McCarthy's eccentricity of riding a horse into the lavish Shamrock hotel he built in Houston seemed a throwback to the antics of the exceptionally rich at the beginning of the century.

When James M. West, Houston oilman and rancher, died in 1958, newspapers discussed his estate of $100 million;

$290,000 of it was in silver dollars cached in his basement. Although West's favorite pastime was riding night patrol with Houston policemen, he left some other assets that drew mournful comment from Lucius Beebe in his Virginia City, Nevada, *Territorial Enterprise:* "Forty-one Cadillacs. Not a Rolls or Bentley or Jaguar or Thunderbird or any car of real distinction or character among them. . . . If the miserable man had owned forty-one Derby winners or maintained forty-one mistresses, he would have achieved some measure of worldly satisfaction in his wealth; some justification for having been around. As it is, no one need envy him. All he rated in the end was a Cadillac hearse."

For every James West there are and were a dozen oilmen whose fortunes have been poured into libraries, art museums, universities, hospitals, and other philanthropies of amazing variety and vast scale. Of the great explorers, only Edward L. Doheny and E. W. Marland spent their fortunes on the lavish and sybaritic scale of the industrial buccaneers of an earlier day. They made their money early and as young men were still under the influence of the glittering example of the old order. Doheny and Marland believed, as did many of their contemporaries, that the privilege of wealth was to display it in the form of yachts, private railroad cars, palaces, art objects, and elaborate parties.

However, Doheny and Marland also belonged to the new era in American thinking that arrived through the awakening of the nation's social conscience. They were as lavish in helping others as they were in providing for their own pleasure. The great difference between them and the rich aristocracy they seemed to copy was that their philanthropies were a part of their living and not a codicil to their wills.

The desire to share their wealth during their lifetime is a more predominant trait among oil explorers than any other group of men who have derived their fortunes from a single source. Almost without exception, each of the explorers started with nothing, and their efforts to succeed were hounded by failures and struggles. They knew what it was to be penniless, to have to borrow money to finance their hopes and the grocery bill. Succeeding, they were keenly aware of the great element of chance involved. Unlike the robber barons, they did not succeed at someone else's expense. The explorers, as one writer observed admiringly about Charles Canfield, "left no human wrecks along the road to success." They were influenced by the realization that they had created new wealth. The majority of them believed strongly that the billions of barrels of oil that poured out from the earth did not belong to them exclusively by the law of capture.

Before Hugh Roy Cullen died in 1957, he gave away cash and oil properties valued at $175 million—93 per cent of his fortune—to universities and hospitals. During one forty-eight-hour period he gave Houston's four hospitals more than a million dollars apiece, with no strings attached as to how it should be spent. Cullen's donations built the University of Houston, which was established for the children of working people who could not go away to college. Roy Cullen had only a third-grade education, but his grandfather, Ezekiel Cullen, was the founder of the Texas school system in the early days of the Republic of Texas.

The $160-million Cullen Foundation, established in 1947, is exceeded in size only by the Rockefeller and Ford Foundations, and the capitalization is the largest amount of money ever given by any single family during the lifetime of the donors. Roy Cullen maintained this was the only way a man could enjoy his wealth while he lived.

"Giving away money is no particular credit to me," he said. "Most of it came out of the ground—and while I found the oil in the ground, I didn't put it there. I've got a lot more than Lillie and I and our children and grandchildren can use. I don't think I deserve any great credit for using it to help people. It's easier for me to give a million dollars now than it was to give five dollars to the Salvation Army twenty-five years ago."

Cullen had occasion to remember that five-dollar donation. It was his first charity and the check was returned marked "insufficient funds."

The multimillion-dollar foundation Mike Benedum and his wife established as a memorial to their only son, Claude Worthington Benedum, who was killed in World War I, is primarily devoted to education, providing scholarships and student-loan funds in twenty-six colleges in West Virginia, western Pennsylvania, and eastern Ohio.

In sharing his wealth during his lifetime, Mike Benedum specialized in projects that called for his active, creative participation. The restoration of his home town of Bridgeport, West Virginia, is a case in point. During the depression, he wanted to help the struggling community, which lies in the bluegrass region and is dependent on cattle raising. The town's three cemeteries, which adjoined each other, were in a deplorable condition, so to provide employment and as a community service he restored them. There were jobs for hundreds of workers, building a road, strong fences, a chapel, and planting 10,000 new plants.

Next, he built a Methodist church of such outstanding

beauty that an average of 12,000 visitors from every state and many foreign nations visit it annually.

His third project was a civic center. Benedum bought the old house in which he had been born, also the adjoining properties. Razing the old buildings, he built a greatly enlarged replica of the old family home to provide libraries, museum, auditorium, dining hall, and club rooms. On the grounds he built a swimming pool, wading pool, an ice rink of the same proportions as the one in New York's Rockefeller Plaza, badminton and shuffleboard courts, and a children's playground.

The result is a striking demonstration of what can be achieved by planned emphasis on cultural development alone. With no new industry, Bridgeport's 1,000-odd population has grown to 5,000. Its property valuations and bank deposits have increased proportionately. Municipal officials and ministers report that the problem of juvenile delinquency is nonexistent. The Civic Center, with its dances and other entertainments, is the nightly rendezvous of all the young people.

Mike Benedum particularly enjoyed sharing the opportunity to make money with those who worked with and for him. Whenever he organized a new exploration company, he reserved a certain amount of stock for his employees to buy if they wished, and financed them to buy it. As a result, messenger boys, file clerks, secretaries, and accountants have profited tremendously. When his secretary, Margaret E. Davis, died in 1957 after working for him thirty-nine years, she left an estate of more than a million dollars.

Waite Phillips followed his older brother to Bartlesville, Oklahoma, from Iowa in 1906 and worked with him for a few years in the booming new oil business. Before Frank

Phillips formed the Phillips Petroleum Company, Waite prospected on his own and rapidly built up a fortune that equaled, or even exceeded, his brother's.

Waite Phillips created an empire for the nation's youth rather than himself. He bought a third of a million acres near Cimarron, New Mexico, for a cattle ranch. It included some of the most wildly beautiful mountain and mesa land in the West, part of the immense land grant acquired by Lucien Maxwell, Baron of the West, with whom Kit Carson made his home and established a trading post where the Sante Fe trail crossed the Rayado River. Phillips built a chain of fishing and hunting camps through the mountainous back country. In 1938 he sold some of the rangeland and presented 127,000 acres, including the spacious ranch house, the chain of camps, the cattle, the buffalo and antelope herds, to the Boy Scouts of America. Philmont Scout Ranch is the largest camp of its kind in the world; annually some 8,000 Explorer Scouts revel in its beauties, its wilderness, and unequaled opportunities for training. Being a practical man, Phillips also presented the Boy Scouts with the Philtower, a 22-story Tulsa office building. It provides income to run the ranch.

Waite Phillips gave the city of Tulsa his million-dollar Italian Renasissance home and gardens, Philbrook, to be used as an art museum with his own outstanding collection to start it. After moving to California, he made another great fortune in real estate and became as active as when he was hunting for oil in Oklahoma.

Tom Gilcrease was one-eighth Creek Indian. His Scotch-Irish father had a farm and small store in Creek Indian Territory and the boy, who was born in 1890, grew up in an atmosphere only his generation knew. The region was park-like, without weeds or underbrush. Only Indian pony trails and sometimes faint wagon tracks marked it. The streams were clear; there was an abundance of deer, turkey, and prairie chicken. Tom Gilcrease's Indian heritage gave him an instinctive love of beauty and a mystic bond with its expression, whether by nature or man. Incredibly, and in the manner of a hero of a romantic novel, this shy lad became an oil explorer and gave to the people the finest collection of American art, manuscripts, and artifacts ever assembled.

When the Creeks were forced to accept allotment of their lands, Tom Gilcrease was alloted a quarter-section in what soon proved to be Oklahoma's first giant oil field, Glenn Pool. The naive boy lost part of his allotment, as so many others did, because he was not equipped to fend for himself when it came to business. Then the Scotch-Irish side of his nature asserted itself. Using some of his remaining royalty income, he bought a drilling rig and began to search for oil. He traveled by buckboard and foot over Oklahoma, studying rocks, trees, and countryside wherever there was oil, getting the feel of it. In areas where there was seemingly no oil, he applied his self-observed geology to picking possibilities. The young wildcatter discovered a series of small fields.

Tom Gilcrease's company prospered. He went into the unexplored Seminole country and wanted to lease all of it, but the geologist he hired talked him out of it, leaving its giant fields to be discovered by others. However, his wildcatting carried him steadily south until he was wildcatting in Texas when Dad Joiner's well came in. Gilcrease took leases that seemed far away from the new East Texas field. He was unable to interest any American oilman in drilling with him, but on a trip to Paris he found an assortment of willing French capitalists. The leases proved to be in the heart of the giant field.

To Tom Gilcrease, great wealth meant freedom to expand the collection of paintings and books that he had started twenty years earlier and which already overflowed his home in the Osage hills near Tulsa. His hobby now became his obsession. As fast as his oil flowed from the ground, he transformed it into the art and culture of the Americas. He was a self-taught collector just as he was a self-taught explorer, but with unerring instinct, he found and bought priceless pieces for what he now knew would be a museum, and the only one of its kind. As an Indian and a Euro-American, he began to build the history of America as its artists, artisans, writers, and makers of history portrayed it from the archeologists' discoveries of the most ancient Indian cultures in North and Central America, through the conquest of the Americas by the white man, the emergence of the United States, to the opening of the West and the disappearance of the Indians. It was a magnificent conception, seemingly impossible for one man to realize, but Tom Gilcrease carried it off. The art world was not aware of what the modest, self-effacing oilman had done until, in 1954, he presented his $12-million collection to the city of Tulsa.

The professional staffs of the nation's big museums were astonished when they learned that among his five thousand paintings, thirty thousand books, thirty thousand manuscripts and documents, ten thousand sculptures, potteries, and artifacts, there was the original commission of General Warren ordering Paul Revere to make his famous ride; the only extant copy of the Declaration of Independence signed by its framers, which Benjamin Franklin had presented to the King of Prussia; the original signed proclamation of Cortez that he had captured Mexico City; the original manuscript of Martin Waldseemüller, the German cosmographer who named America; masterpieces of all the great American painters; and the rarest collection of Indian art in the nation.

Tom Gilcrease couldn't get his oil out of the ground fast enough to pay for the items he wanted and had under option. Tulsans voted a bond issue to provide two and a quarter million dollars to buy them, and Gilcrease dedicated the equivalent amount of oil from his reserves to repay it.

"If some boy or girl out of the thousands who come to this museum may get one inspiration or write one book or have one idea that may help others, then what I have done will be worthwhile," Tom Gilcrease said.

What Edward Doheny, E. W. Marland, Mike Benedum, Waite Phillips, H. R. Cullen, and Tom Gilcrease did with their wealth has been repeated in less spectacular form in every community where successful oil explorers have lived.

Even the fortunes not derived primarily from exploration have been responsible for some of the nation's greatest philanthropies. John D. Rockefeller devoted as much time to distributing his immense fortune as he did in acquiring it, his gifts totaling more than half a billion dollars. His son and grandchildren, whose individual fortunes account for seven of America's biggest, have literally devoted their lives to philanthropy.

The Mellon family foundations have channeled more than a quarter of a billion dollars into as great a variety of philanthropic projects as the Rockefellers'. Although the Mellon fortunes have other sources in addition to oil, Gulf Oil Corporation has been the star jewel in the family crown.

The Pew family of Philadelphia, whose Sun Oil Company grew from Spindletop days to the fourteenth largest oil company, primarily through its refining and marketing activities, established a $170-million foundation.

Of even greater importance to the nation than this sharing of oil wealth has been the sharing of ideas by the great exploration thinkers to assure the nation of a continuing abundance of its most vital and elusive natural resource.

DeGolyer, Wallace Pratt, Lewis Weeks, and A. I. Levorsen dedicated their lives to developing and establishing a philosophy of exploration to guide and stimulate new generations. They turned muddle-thinking into simple, crystal-clear precepts and concepts.

Everette DeGolyer was the first to make explorers aware that prospecting is not one long, slowly developing art, but a series of separate techniques exhausting themselves through use. As the father of petroleum geophysics, he proved this principle as well as spelling it out in scientific papers and lectures. DeGolyer hammered home the point that oil is discovered as the result of multiple thinking and multiple effort, the greatest single element in all prospecting, past, present, and future, being the man willing to take a chance. He constantly exhorted geologists to practice "loose thinking" in order to stimulate new ideas.

The man was accomplished and zestfully creative in a dozen fields. He loved people, literature, history, art, music, writing, conversation, correspondence, business, economics, and science with almost equal intensity, aware of their importance and contributions to the art of living. He built his magnificent Spanish-style home in Dallas around a large library containing 20,000 volumes, the finest private collection in the world on the Southwest. The DeGolyer Foundation will eventually make it a research library open to scholars.

Books meant so much to DeGolyer that once in Chicago, on finding a rare copy of *Three Years Among the Mexicans and Indians* by Gen. Thomas James, published in Waterloo, Illinois, in 1846, he asked the book dealer to mail it to him in Dallas.

"But it's such a small volume, why don't you take it with you?" the dealer asked.

"Oh, no!" DeGolyer said in a shocked voice. "I'm going by plane and I wouldn't want to take a chance on anything happening to the book."

DeGolyer collected from all over the world 15,000 manuscripts and books on the history of science, presenting them to the University of Oklahoma. He gave a $150,000 collection of modern first editions of poetry, novels, and philosophy to the University of Texas. When the *Saturday Review of Literature* was about to go bankrupt, he financed it until it was out of trouble and then, at no profit, sold back the stock he had purchased, in order that the magazine's editors could own it as well as publish it.

DeGolyer's preoccupation with literature was one of the reasons he was able to communicate his scientific thoughts so brilliantly and effectively. At the time of his death in 1956, DeGolyer, then an advisor to the Atomic Energy Commission, wrote his friend Wallace Pratt that he was planning to call a conference of America's greatest thinkers in all fields of science to see if combined brain power outside the oil industry could, by "loose thinking," evolve the greatest of all oil finding techniques—the one which would find oil directly.

DeGolyer felt that Pratt, "more than any other man, has brought to the general public an awareness and appreciation of the importance of the science of geology to the arts of prospecting for and producing oil. In his own words, 'The winning of oil is a geologic enterprise.'"

Wallace Pratt indelibly stamped the thinking of explorers with his awareness that "oil in the earth is far more abundant and far more widely distributed than is generally realized. The prime requisite to success is freedom to explore."

Lewis Weeks was concerned with the great horizons of world oil. As senior research geologist and then chief geologist of Jersey Standard, his monumental studies of where and how oil occurs are basic texts in colleges and universities throughout the nation. He was the only geologist who studied the habitat of oil in every world basin, and the principles he developed are a foundation and inspiration for the discovery and development of world oil resources.

A. I. Levorsen taught the simple, yet profound truth that any undiscovered oil or gas fields exists only as an idea in the mind of the explorer. He devoted his time to being the champion of unorthodox thinking and in his papers, books, and lectures pointed out specific areas in America that should prove to be new oil provinces when this sort of thinking is correctly applied.

Levorsen was the first man to make geologists aware of the great importance of stratigraphic traps, which have no structure but are simply changes in the rock, and can be expected to contain the bulk of our undiscovered oil. The most difficult and challenging of all pools to find, stratigraphic traps require the most unorthodox thinking of all.

Never has an inexact science had such brilliant, articulate thinkers to stab others awake in perfecting the art of prospecting for oil in America and throughout the world.

After a hundred years of oil America led the world's oil industry. The men who had made it so had done more to change human affairs and human behavior than any other events since Edwin Drake had drilled the first commercial well.

5: *The eagle folds its wings*

As the 1950's came to a close, the ranks of the great wild-catters and scientists had thinned. The industry they had pioneered was beginning to change in profound and disturbing ways. National and international political developments were conspiring to end the abundance of cheap energy in America. The stage was being set for a series of puzzling crises.

The principal villain of the unfolding energy scenario was a little-publicized, seemingly dull Supreme Court decision in 1954. The court ruled that natural gas sold in interstate commerce was subject to Federal Power Commission price controls. Natural gas had essentially been a by-product of oil production. Vast quantities were flared until after World War II, when it became the glamor new energy source. It was the cleanest, most efficient, and environmentally most desirable of all the fossil fuels. Now that artificially low price regulations made it also the most inexpensive, demand skyrocketed. Soon, interstate pipelines were purchasing 70 per cent of the nation's production. Natural gas provided one-third of all energy and supplied nearly half of the energy for industry and the residential market. Its cheapness depressed the market price of competitive fuels—coal and fuel oil. This forged the first powerful link in the chain of events leading to a lopsided, unbalanced development of energy resources and the buildup of increasing dependence on foreign oil.

Because of competition from natural gas, refiners found that fuel oil was selling at the refinery for less than the cost of the crude oil from which it was made. So they converted more of each barrel to lighter products—gasoline, naptha, kerosene, and diesel oil—on which they could make a profit. New refineries

335

were designed to make these higher-priced products. Consequently, as the demand for fuel oil increased, it had to be met by imports. It was not long before domestic production of fuel oil was meeting only 30 per cent of the nation's requirements. The East Coast, which had the heaviest consumption, ultimately became dependent on imports for almost 94 per cent of its fuel oil.

Another new source of competition was a flood of cheap Middle East crude oil. The shape of things to come in world oil resources had been accurately predicted by E. DeGolyer during World War II. When war broke out in 1939, Saudi Arabia, where oil had been discovered two years before, was producing only 10,000 barrels a day. However, the oil was so strategically located and urgently needed that the U.S. government sent a mission, headed by DeGolyer, to estimate proved and probable oil reserves in the Middle East. DeGolyer reached a startling conclusion. "The center of gravity of world oil production is shifting from the Gulf-Caribbean areas to the Middle East, to the Persian Gulf Area, and is likely to continue to shift until it is firmly established in that area," he said.

At war's end, international oil companies rapidly began proving the correctness of his insight. As they developed huge reserves, foreign oil became cheaper than domestic oil. This was primarily due to the difference in production costs. An average U.S. well produces about 16 barrels a day and generally needs to be pumped, while the average Middle East well flows 5,000 barrels daily with no production assistance necessary. American oil imports began to rise at such an alarming rate that in 1959 President Dwight D. Eisenhower restricted imports for reasons of national security. His purpose was to encourage increased domestic production.

However, stemming the flow of imports was not enough.

Like Job, the oil industry was plagued by a variety of troubles. Although few realized it at the time, the efforts of the American oil explorers at home peaked in 1956. That year almost 9,000 new field wildcats were drilled, but exploration began to decline so precipitously that 15 years later only 4,500 wildcats were sunk.

Imports and natural gas controls, which had decreased exploration incentives, were not the only reasons for the abrupt exploration drop. The industry was caught in a cost price squeeze. Between 1959 and 1969, wages in the domestic oil and gas industry increased 40.8 per cent, oil field machinery was up 24 per cent and oil well pipe up 17 per cent. By contrast, the price of domestic crude oil was only 3.7 per cent higher. This reduced the amount of risk capital available for exploration, particularly for the independents.

At the same time, the emphasis of major company exploration shifted from onshore to offshore to look for big fields. This sharply reduced the amount of capital assistance major companies traditionally provided independents in the form of contributions to the cost of drilling wildcats in areas where the major companies also held lease acreage. Furthermore, billions of dollars paid by major companies as bonuses in order to obtain state and federal offshore leases meant that much less exploration money to put into the ground.

Adding to the industry's exploration woes was a change in the tax laws in 1969 which reduced tax deductions, designed to encourage oil and gas discovery and development, and raised the industry tax bill by $600 million annually. This was particularly tough on the indispensable independent oilman.

The sum total of all these economic and political disincentives dried up risk capital at a time when the industry needed it most in order to find new reserves to meet soaring demand. All

the decrease in domestic petroleum exploration and development resulted from the decrease in independent oilmen's activities. In 1956 independents spent $2.5 billion for domestic
exploration and development. By 1971 they spent less than $1.2
billion. Major oil company domestic exploration and development expenditures remained constant, excluding lease bonuses
to federal and state governments. During the same period the
demand for oil doubled, and consumption of natural gas increased 120 per cent. Oil and gas were now supplying three-
fourths of the nation's energy needs. The most perturbing
statistic of all was that the number of America's independent
oilmen had been reduced in 15 years from 20,000 to 10,000.
Politicians and consumers alike turned a deaf ear to the dis-
tressed warnings of independents that the nation was dis-
mantling its exploration industry and was headed for serious
domestic oil and gas shortages.

But in 1968 America flexed its oil muscles once again.

For centuries, the Eskimos had known about oil seeps on the
North Slope of Alaska bordering the Arctic Ocean. Following
World War I, geological exploration indicated that the area
had real oil potential, and President Warren G. Harding established Naval Petroleum Reserve Number 4 to preserve resources for a future national crisis. During World War II,
President Franklin D. Roosevelt dispatched an expedition to
explore for oil. A field containing an estimated 70 million barrels
of oil was discovered, but the hostile environment and the
difficulty of transporting Arctic oil to where it could be used
discouraged serious development.

In 1964, the presence of oil, development of Arctic technology, and increased demand stimulated major oil companies
to lease state lands on the North Slope and continue exploratory
drilling. Atlantic Richfield and its partner, Humble Oil and

Refining (which would soon change its name to Exxon), drilled 10 dry holes at a cost of millions of dollars. In the winter of 1967 they decided to drill one last-shot well before abandoning the area and picked a remote site at Prudhoe Bay. They began to drill where Wallace Pratt had recommended that Humble explore more than 20 years before. In April, 1968, at a little more than 12,000 feet under the frozen earth, they hit oil. The well was capable of flowing 2,400 barrels a day, but it was not until they drilled a second big producer seven miles to the southeast that they realized they had found a giant oil field.

As other companies rushed to drill their leases, the astonishing truth unfolded. Prudhoe Bay was a giant among giants—America's biggest oil field, containing a conservatively estimated 10 to 15 billion barrels of oil and 26 trillion cubic feet of gas, covering 250,000 acres. This was two or three times as much as the fabled East Texas field which had held the record. The "last-shot" gamble had found reserves representing 29 per cent of America's crude oil and about 11 per cent of its natural gas. A score of major companies promptly bid $900 million for additional state leases to explore what promised to be one of the world's great oil provinces.

The immense problem of how to bring this huge new source of energy to the "lower 48," where it was so greatly needed, was confidently tackled by eight companies owning Prudhoe Bay reserves. In 1969 they decided to form the Alyeska Pipeline Service Company to engineer and build a pipeline, with 2 million barrels daily capacity, from the North Slope 800 miles through Alaska's rugged terrain to the icefree Pacific Ocean port of Valdez. They expected to complete this unprecedented engineering feat by 1973 and imported a $300,000,000 mountain of iron pipe. But their optimism was soon shattered by unexpected and unrelated developments in California.

In 1968, the federal government invited bids for oil and gas leases on its deep offshore tracts beyond the three-mile limit in the Santa Barbara Channel, off the coast of southern California. Oil companies leaped at the opportunity. Offshore drilling technology had developed sufficiently to explore these deep waters which the successful exploration of state tidelands had indicated might be the habitat of giant oil fields. After keen, competitive bidding, a large number of companies paid $603 million in bonuses to drill in the Channel.

By January, 1969, five offshore rigs were in place in the federal tidelands $5\frac{1}{2}$ miles from the beaches of the famed resort community. On January 28th a catastrophe occurred on a Union Oil Company well drilling on a tract owned jointly with Mobil, Texaco and Gulf. The crew was routinely running the drill up out of the well to change bits when there was a huge gas blowout. They managed to get the drill pipe back in the hole and a blowout preventer rammed shut. But the gas pressure was so strong that it forced its way through the well wall into a fault zone of cracks and fissures leading to the Channel floor. Soon six-foot-high gas bubbles were erupting in the water, and later in the day a big oil slick began to form. It took eleven days of prodigious effort to kill the well and plug the leak by pumping thousand of tons of cement slurry down into the well. But before this was accomplished, an estimated 10,000 barrels of oil had spilled into the Channel, polluting beaches and over 200 square miles of ocean.

The ill-fated well had, indeed, discovered a giant oil field—Dos Cuadros, containing 221 million barrels. But more than oil erupted at Santa Barbara. Intense press, television, and national magazine coverage presented the event as a major ecological disaster. The sight of dead birds, fouled beaches and predictions of permanent damage inflamed public opinion against the oil companies. The Secretary of Interior immediately suspended activity on 35 of 69 Santa Barbara leases and ordered a leasing moratorium for all U.S. offshore areas. California banned new drilling on all state offshore leases.

The beaches were clean four months later, following a massive campaign launched by the companies. But there was no way to clean up the psychological damage that had been done to the public mind concerning environmental dangers they now believed were inherent in offshore oil production. A new disease had been discovered—petrophobia.

Scant attention was paid to an independent study by the University of Southern California's Allan Hancock Foundation, involving about 40 scientists, which was conducted over 18 months following the spill. They found that the bird population had not been decimated, damage to sea life was not widespread, the area recovered environmentally, and all animals were reproducing normally. The Channel fish catch was greater in a six-month period after the oil spill than in a comparable period the year before.

Little publicity was given to the fact that there were 16,000 U.S. offshore producing oil and gas wells, most of them in the Gulf of Mexico. Including Santa Barbara, there had been only four major oil spills, in excess of 5,000 barrels, in 25 years of offshore operations. Two of the the others resulted from fires, and one from a storm. Nor was it generally recognized that Nature was the biggest oil polluter in the Santa Barbara Channel. Over the years, in the same area, natural oil seeps have leaked more oil into the Channel each year than did the Union well blow-out. Indeed, the first search for California oil in the 1860's was stimulated by the report on oil seeps on land and in the Channel which was made by Benjamin Silliman, the Yale University professor whose report on Pennsylvania's rock oil had inspired investors to drill America's first commercial oil discovery in 1859.

The real significance of the Santa Barbara oil spill was that the dramatic occurrence ignited the fuse of an uneasy American conscience. While the average American was not yet aware that the nation had an energy problem, *everybody* knew there was an environmental crisis. Like the energy problem, it had been long in the making, and the two were interrelated. Concern over what was happening to the environment as the result of new technology, with its outpouring of goods and chemical products and tremendous consumption of resources, had been increasing steadily since the early 1960's.

Three-fourths of America's population were living in congested metropolitan areas, and more than half of all Americans were living within 50 miles of the shores of the nation's oceans and Great Lakes. This concentration, combined with greatly increased industrialization and a new era of "throw-away" products, had accelerated air, water, and land pollution. "Prophets of Doom" were warning that mankind would quickly destroy itself and exhaust the earth's resources. An alarmed public began to clamor for social action.

Congress had passed a Federal Water Pollution Control Act in 1965 and an Air Quality Act in 1967, but little headway had been made in cleaning up either water or air. Consequently, Santa Barbara became the banner of an environmental crusade which quickly exploded into a dominant political issue with 1969 becoming the Year of Environmental Awakening.

Government was spurred into action. By the end of the year Congress passed the National Environmental Policy Act, which made it national policy to "maintain conditions under which man and nature can exist in productive harmony, and fulfill the social, economic, and other requirements of present and future generations of Americans." It directed that the policies, regulations, and public laws of the United States "shall be interpreted and administered in accordance with the policies set forth in this act." It set the requirement for all federal agencies to prepare a detailed statement of justification of "major federal actions significantly affecting the quality of the human environment." This was called an "Environmental Impact Statement." In effect, the provision meant that interested citizens can challenge in the courts any federal agency's ruling on a matter "affecting the quality of the human environment" on the ground that the ruling does not comply with the intent of the environmental act. It gave environmental action groups a powerful weapon to use in every area affecting energy development. Especially, since 50 per cent of the nation's remaining oil and gas potential, 40 per cent of its coal, 50 per cent of its oil shale and 60 per cent of its geothermal energy sources are located on federal land. Furthermore, all the activities of energy industries are affected by state and federal legislation.

In April, 1970, environmentalists fired their first shot under the new Act. It was a broadside. It stopped the building of the Alaska pipeline. Three conservation groups—the Wilderness Society, Friends of the Earth, and the Environmental Defense Fund—charged in court that the Interior Department had failed to make an adequate environmental impact review of the project. They claimed that the pipeline, in traversing 400 miles of frozen tundra, three rugged mountain ranges, seventy streams and river, and one earthquake zone, would irreparably damage "the last great wilderness in the United States." They claimed it would adversely affect plant and animal life and result in disastrous oil spills caused by earthquakes and tanker accidents. A federal court granted them a temporary injunction enjoining the Secretary of Interior from issuing a construction permit. The suit would delay beginning pipeline construction for five years and escalate the estimated cost from $1 billion to $9 billion.

During the first three years of the National Environmental Protection Act, so many lawsuits were initiated by private groups that federal agencies were compelled to prepare 3,600 Environmental Impact Statements. Environmental political action literally stopped energy development dead in its tracks. Environmentalists put political roadblocks in the way of all action and planning for energy growth. Their targets were not only the Alaska pipeline and offshore oil, but building of oil refineries and nuclear power plants, development of oil shale, and strip coal mines.

Meanwhile, during the same period, the nation went from one energy shortage cliffhanger to another.

In the summer of 1970 the public was bewildered when the federal government announced it had alerted its Emergency Petroleum Supply Committee to be ready for a possible world oil emergency. Newspapers from coast to coast carried front-page stories that there would be a winter fuel oil crisis. They warned of shortages, higher fuel prices, industrial shutdowns, worker layoffs.

It seemed implausible that such dire predictions could be triggered by the damaging of a Middle East pipeline and a cutback in Libyan oil production. But such was the case. A bulldozer in Syria accidentally cut Tapline, a pipeline carrying half a million barrels of oil daily from Saudi Arabian oilfields to the Mediterranean and crossing through Jordan, Syria, and Lebanon. The Syrian government refused to let repairs be made until higher fees were paid—an argument which would last for nine months. Consequently, Saudi oil had to be shipped the long way around Africa to European and American markets, since the Suez Canal had been closed during the 1967 Arab-Israeli war. A world tanker shortage developed. At the same time, the government of Libya, the free world's fifth-largest oil-producing country, cut back American oil companies' production by 800,000 barrels daily in a power play to force them to raise prices and the taxes paid to Libya. This cutback intensified the tanker shortage. Additional Middle East production, an attempt to make up the Libyan loss, had to make the long haul. The two events disrupted the entire world oil supply and threw consuming countries into a turmoil.

The vulnerability of the United States resulted from its energy imbalance, environmental restrictions, and the effects of natural gas price controls. Because of air quality standards, power plants had converted rapidly from high-sulfur coal to oil and were competing for low-sulfur oil with commercial and individual customers. Most of this low-sulfur oil came from North Africa, either as crude oil imported to be blended with high-sulfur domestic and imported oil or as imported fuel oil from European refineries using North African crude oil.

The heavily industrialized East Coast was dependent on imports for 94 per cent of its fuel oil, and fuel oil supplied 45 per cent of the total energy it consumed for commercial and industrial purposes. True to predictions, natural gas, as an alternative to fuel oil, was now in short supply. Price controls had reduced incentives to look for new fields, and reserves had declined for the second straight year. Gas companies were compelled to refuse new customers and alert established users that their supplies would be reduced. Consequently, oil imports would have to be increased in order to meet rising East Coast requirements. However, the closing of Tapline, the Libyan cutback, and the world tanker shortage meant that little fuel oil from European refineries or imported low-sulfur crude oil would be available to the East Coast.

As it finally turned out, people on the East Coast didn't go cold, and there were no industrial shutdowns or worker layoffs. But fuel oil prices did go up. The highly-publicized fuel oil shortage of 1970–71 was averted by industry and government emergency measures. The nation's reserve crude production capacity had dwindled to about one million barrels daily, but producers drew on this. Refiners, with the incentive of increased fuel oil prices, changed their product output and made up a great part of the shortage. The government relaxed restrictions on imported heating oil and crude from Canada. In January, 1971, Tapline was reopened. As fuel flowed to the Mediterranean, the tanker shortage was alleviated. Prices went down.

Those in industry and government who warned that this emergency demonstrated the appalling vacuum in which America's non-energy policy was operating were accused of crying wolf. Congressmen and the media charged oil companies with having conspired to create "artificial shortages" in order to raise prices and eliminate independents. However, with ample current supplies, the public soon lost interest.

During 1972 the rate of consumption increased more than twice as fast as in 1971. Domestic production declined and imports soared. European demand had also been increasing steeply, and oil became in tight supply worldwide. The wolf came back to the American door in the winter of 1972–73, and this time he brought the whole pack with him. Fuel shortages were severe enough to hit every part of the country in one form or another. Workers were laid off when plants and factories couldn't get enough gas or heating oil. Schools and universities closed for lack of heat. Major airlines were forced to ration jet

fuel at Kennedy airport. Furthermore, refiners predicted that
there would be a critical spring and summer gasoline shortage,
with rationing becoming a possibility.

When Congress convened in January, 1973, it responded
quickly to public anger. A flood of more than 1,000 bills was
introduced by Congressmen from almost every state. For the
first time, federal officials began saying publicly that "energy
is our Number One domestic problem." President Richard M.
Nixon announced plans to form a Council on Energy Policy.
He delivered a major energy message to Congress urging it to
pass legislation to provide incentives to increase domestic
energy supplies. He abolished import quotas, replacing them
with a tariff system, and ordered a triple increase in the amount
of federal offshore acreage to be offered for oil exploration
leases.

However, Congress had its own notions. It busied itself more
with trying to fix the blame for shortages than in considering
what should be done to prevent continuing shortages. Confusion surrounding the energy crisis was deepened during the
summer of 1973. By May the nation seemed headed for a gasoline shortage of major proportions, for gasoline demand had
spurted ahead of supply at a record rate. Supplies were so short
that more than 1,000 service stations were closed by Memorial
Day. Congressmen were besieged with complaints from independent station owners and marketers that they could not obtain supplies. Major company suppliers cooperated with the
government in a voluntary allocation program seeking to divide
available supplies equitably among all consumers. In June the
oil industry managed to turn things around. Refiners ran their
plants flat out. They operated at levels believed to be impossible. They recommissioned obsolete equipment. In addition, the
new import policy permitted foreign crude oil and gasoline to

pour into the country. All these measures increased prices about two cents a gallon.

The shortage had been stopped, but it didn't solve the problem. It actually made it more difficult to understand. The oil industry was attacked more severely than ever before on the now standard charges that major companies had rigged the shortage to raise prices and put independents out of business.

Congress had not been entirely inactive in regard to energy legislation in the midst of the multitude of contentious hearings attempting to fix the blame. After five years of litigation and completion of the most exhaustive environmental impact studies ever carried out in history, both the Senate and the House voted in September, 1973, to exempt the trans-Alaska pipeline from further court review under the National Environmental Policy Act of 1969. Although it would be four years before the pipeline would be completed, the nation's greatest oil discovery could eventually begin supplying the domestic oil for which the nation was so thirsty.

The decision had come too late to help prevent what was about to happen—events that would change world oil supply and energy economics in such a drastic fashion that they would never be the same again.

The Libyan cutback in 1970, which had helped to precipitate the first major U.S. potential energy crisis, sparked a Middle East oil revolution. For ten years a cartel of Arab oil countries, Iran, and Venezuela, known as the Organization of Petroleum Exporting Countries, or OPEC, had been struggling to wrest control of their oil resources from the international oil companies to which they had given development concessions. These concessions were on a fifty-fifty profit-sharing basis, but the countries wanted complete ownership and complete control of world oil prices.

Libya's power play had succeeded in increasing its government profits to 55 per cent and raising oil prices by 30 cents a barrel. Venezuela, the principal supplier of U.S. imported crude oil, promptly increased taxes so that the government would receive 80 per cent of profits, and its prices rose. An OPEC battle to achieve its objectives was now in full swing. Throughout 1971, 1972, and most of 1973, the oil companies lost every round in complicated participation, tax, and price negotiations and by unilateral government action.

By September, 1973, a series of astounding changes had occurred. Algeria nationalized 51 per cent of oil companies' operations. Iraq nationalized 100 per cent of oilfields containing 60 per cent of its production and reduced oil companies' holdings to a total of 300 square miles. Saudi Arabia, Kuwait, Abu Dhabi, and Qatar forced oil companies to agree to 25 per cent government participation in their operations. Iran, which had nationalized oil in 1951 but had concluded a fifty-fifty profit-sharing agreement for operations by a consortium of international oil companies, announced a complete takeover of the consortium, with oil companies to act as customers, not producers. In addition to these nationalizations, participations, and the takeover, OPEC had also achieved a 12 per cent boost in oil prices. But this was just the beginning.

In the United States all these events were as far removed from the American consumer's consciousness as though they had been occurring untelevised on another planet. The American "energy crisis" which had surfaced in 1973 as yet had no understandable connection with Middle East oil affairs. Besides, even the gasoline shortage had been run off the front page in the summer of 1973 by the beginning of the nationally televised Senate Watergate hearings.

President Nixon deliberately had not mentioned in his

national energy message that government predictions indicated that the United States would be importing 55 per cent of its oil needs by 1985, unless national energy policies changed. And the oil would be coming primarily from Arab oil producing countries in the politically unstable Middle East, where a renewed Arab-Israeli conflict was heating up. With Watergate investigations underway, his advisors felt it unwise to invite controversy over any reason to change the U.S. pro-Israel policy. Consequently, the most important, little-publicized event in 1973 was the unsuccessful effort of King Faisal of Saudi Arabia to persuade official Washington that his country, which had the largest production and reserves, would find it difficult to raise its oil production to meet projected American needs unless the United States used its influence with Israel to bring about a political settlement in the Middle East satisfactory to the Arabs.

The Arab oil countries had been studying American energy surveys more closely than Congress. They knew the extent of the energy crisis in which the United States was already involved and how long it would take to solve it. They also knew that only the Middle East and North Africa, suppliers of 90 per cent of international oil and owners of three-fourths of the world's proved oil reserves, could meet the soaring oil demands of consuming countries. The United States was importing 37 per cent of its oil needs, Western Europe 93 per cent, and Japan almost 100 per cent.

The Arabs' political target was the United States because of its all-out support of Israel. Only 5 per cent of American consumption was supplied by Arab oil, so an embargo could damage, but not wreck, the American economy. However, Arab oil supplied Western Europe with 73 per cent of its needs and Japan with 45 per cent of its consumption. By cutting back oil supplies to these countries, the Arabs could hope to force them to pressure the United States to influence Israel to withdraw from territories occupied in 1967 and make a peace treaty.

By the end of September it was obvious that the United States was in no hurry to use its influence with Israel to get negotiations for peace off dead center. The Arabs decided on the shock treatment of open warfare. On the morning of Yom Kippur, the holiest of Jewish days, when the synagogues were filled with worshippers, Egypt and Syria simultaneously launched a massive surprise invasion against Israel across the Suez Canal and Golan Heights. After two weeks of bloody fighting the Arabs unleashed their ultimate weapon. By cutting back almost a fourth of their oil production, which directly affected Europe and Japan, and totally embargoing oil shipments to the United States, the Netherlands, Portugal, South Africa, and Rhodesia because of their pro-Israel policies, they dropped a political and economic bomb whose spectacular fallout spread rapidly around the world.

Although the United Nations arranged a cease-fire three weeks after Yom Kippur, the international energy war intensified. Three days after the shooting war began, the Arab Gulf countries and Iran unilaterally announced a staggering 70 per cent increase of posted oil prices, amounting to $2 a barrel. Then, two months after the cease-fire, at the end of December, 1973, they jumped the price of oil another 130 per cent. Other OPEC countries followed suit. This meant the price of oil had more than quadrupled within a year, and the oil-consuming nations' import bill in 1974 would rise by an incredible $95 billion over 1973 just for the exporting governments' take in royalties and taxes. This did not include production costs, profit margins, and transportation.

Oil imports into the United States did not begin to dip because of the Arab embargo and cutback until December. However, the first OPEC price jump quickly hit the American consumer. Petroleum product prices had been under government control since August, 1973, but regulations allowed oil companies to pass along increased costs of raw material and purchased products to their customers. Since one out of every three barrels of oil consumed in the United States came from abroad, the costly foreign oil boosted prices.

Initially, Americans adjusted well to conservation measures the government urged voluntarily or set arbitrarily. Gasless Sundays, lowered thermostats, reduced speed limits, going back on daylight saving time, and turning off lights were all manageable things. But when fuel oil and gasoline shortages became realities, gasoline lines formed, and prices rose astronomically as the result of the second OPEC increase, the national mood became ugly, bitter, and accusing.

There was a great outcry when oil companies announced their 1973 fourth-quarter profits. Industry earnings were up 63 per cent over the same quarter in 1972. It was front-page news, and the number one story on every television network. Actually, the earnings percentage increase of chain grocery stores, aluminum and copper companies, and others, exceeded those of the oil companies in that quarter, and the oil industry's return on capital investment was no more than the average of all U.S. industry. But nobody was standing in line to buy food or metals, whereas every American motor vehicle driver was going through one of the most frustrating experiences in his personal history.

The prevalent belief that the crisis resulted from "oil profiteering" and "collusion" between international oil companies and Arab governments was deepened after the oil embargo ended in March, 1974, and the largest oil companies' first quarter profits were up 78 per cent over the same period in 1973. Almost nobody believed the explanations of the oil companies. The 30 largest oil companies reported that 75 per cent of their increased revenues occurred outside the United States and 85 per cent of the profits growth came from abroad. Factors which made oil profits abroad seem higher than they really were included the devaluation of the dollar, inflation, and so-called "profits" which, in fact, were increased book value of their inventories of crude oil to reflect new higher prices. All this was too complicated for the average American to understand when he was now paying 10 to 15 cents more a gallon for gasoline. People simply refused to believe that the oil companies only made an average profit of two cents a gallon on gasoline made from the mix of domestic and foreign oil.

Furthermore, those who had not been following the OPEC script did not realize that in the breakdown of a barrel of imported oil landed in the United States at a price of $11 to $12, $7 went to OPEC governments, 50 cents or less represented profits to the major oil companies in the country where it was produced, and the rest was the cost of handling and tanker charges. Since one-third of the gasoline sold at American pumps was made from imported oil, the major oil company profit on its foreign oil was less than half a cent a gallon.

Domestic crude oil prices also had increased, but not as sharply as imported oil. Since 1971, U.S. production had been under government price and profit margin controls, but they applied only to oil being produced as of then. This portion was called old oil and represented three-fourths of production at the time of the embargo. New discoveries and stripper production—wells producing 10 barrels or less—were non-controlled. As gas lines began to form, the Cost of Living Council authorized a dollar increase in the price of old oil, making it $5.25 a barrel, to provide incentive for discovering new supplies. Because of the shortage of supplies, the price of uncontrolled oil increased by $4 a barrel to $10. So in averaging U.S. production, there was about a 24 per cent overall price increase. Out of a 10 cents per post-embargo gallon of gasoline price rise, about 4 cents was for domestic price increase, 3 cents for imported oil increase, and a 3 cent increase granted by the government to service station dealers and jobbers, the majority of whom were independents and were affected by the reduction in sales volume.

When Congress convened in January, 1974, it was in a punitive mood. It was also running a higher political fever than usual. The Watergate crisis was rapidly moving from investigation to impeachment hearings. The political atmosphere was charged with intense rivalries between presidential aspirants who saw the energy crisis as an ideal opportunity to showcase themselves. There was a blizzard of some 3,000 energy and energy-related bills introduced in the Senate and House.

However, what made the news were the immediate investigations launched on whether or not there really was an energy crisis, who was to blame, and why. "Big Oil" was the target. The industry was on trial. Rhetoric and accusations made the headlines. Facts to disprove them attracted little attention. Oil industry credibility sank to new lows.

Nevertheless, positive action to find new oil resources was underway. Independents had been drowning in the Sargasso Sea of depressed domestic prices. The new high prices were like the arrival of a rescue squad. They began to plow back their profits in the most vigorous exploration program for new oil and gas sources the country had seen in 15 years. In the second quarter of 1974 exploratory gas well completions were 70.5 per cent higher, and exploratory oil well completions 61 per cent higher, than during the same period in 1973. Independents were reworking shut-in and abandoned wells and bringing them back to life, including old wells in Pennsylvania where America's first oil was discovered. They were poking down holes in old, supposedly worn-out areas and finding new oil and gas. Also, major oil companies were investing twice as much in the United States as they were abroad. Their capital expenditures in the United States for 1973 and 1974 were two and a quarter times as large as their profits.

While Congress, the public and the oil industry were trying to cope with the domestic shambles resulting from the Arab embargo and cutbacks, the other consuming countries were in equal disarray. They vociferously blamed the United States for their energy woes and began beating a path to the oil producing countries' doors to make special deals to alleviate their situations. The United States managed to convene in Washington an international energy conference of consuming countries which agreed to try to start a dialogue with producing countries. At the same time, Secretary of State Henry Kissinger began his shuttle diplomacy to try to obtain a genuine settlement of the Arab-Israeli conflict.

The new petropolitical economic problems were the thorniest ever faced by the industrialized nations and the developing countries. The world money flow to OPEC nations was the greatest transfer of wealth in history, and there were no rules in this new game.

Meanwhile the Arab OPEC countries had not been idle. They announced 60 per cent takeovers of foreign oil companies and that negotiations were underway for 100 per cent takeovers. Not only had the center of gravity of world oil production shifted to the Middle East, but it was now the center of political world oil power. The oil companies no longer had any leverage to keep prices down. Symbolically, Saudi Arabia had now become Number One in free world oil production. The United States lost the crown which it had held since the turn of the century. The consumer was the great loser in the swift rush of events. Never again would there be cheap energy.

Throughout 1974, there was no effort by Congress to address itself to the problem of how to increase domestic energy supplies. On the contrary, there were strong efforts to roll back the price of domestic oil, despite the evidence of the new exploration boom because of higher prices. In addition, Congress, mindful of November elections and the national anti-oil-company mood, engaged in a strenuous battle over new tax measures to satisfy voters that Congressional watchdogs would never permit oil companies to profit unwarrantedly at their expense. The battle was not *whether* to tax the oil industry, but how much and by what means, regardless of its effect on exploration and development.

The new Tax Reduction Act was passed and signed in March, 1975. It wiped out the controversial percentage depletion tax allowance—originally designed to provide exploration incentives—for all but the smallest operators and tightened foreign tax credits. Coupled with other "reforms," by increasing taxes it removed approximately $2.5 billion annually from the working capital of the oil industry. The practical result was an immediate cutback in planned exploration and development programs and in drilling rig and seismic activity.

Then, in December, 1975, after two years of inaction and long, acrimonius debate, Congress passed and President Gerald R. Ford reluctantly signed, the first attempt to formulate a national energy policy. For all intents and purposes, the new Energy Policy and Conservation Act, dominantly reflecting liberal ideas, might have been drafted in Saudi Arabia rather than in Washington. It was difficult to see how Arab interests could have been better served. The Act imposed another 40 months of price controls, extended them to new, released, and stripper crude, and rolled back prices on these categories to $11.28 a barrel from free market prices of $12–$13.50. It retained the ceiling of old oil at $5.25 and mandated a composite average of $7.66 per barrel of upper and lower tier prices. It continued the regulation of natural gas at artificially low prices. The Act reduced the petroleum industry's income by another $3 billion annually, with almost half of that loss coming from the pockets of independents, since 40 per cent of the oil they produced was previously exempt from controls as opposed to 30 per cent of major companies' output. New domestic oil and gas exploration suffered another crippling blow. Rather than promoting self-sufficiency, the Act guaranteed increasing dependence on foreign oil imports.

During the 1976 presidential election, imports continued to rise, averaging almost 2 million barrels a day more than in 1973, when the Arab oil embargo was launched. Saudi Arabia took over the lead as the Number One imports source, with imports from that country jumping from 702,000 barrels a day in 1975 to 1.2 million barrels daily in 1976. Domestic consumption increased, and oil and gas reserves continued to decline, although exploratory drilling was higher than it had been since 1970.

In January, 1977, President Jimmy Carter was inaugurated in the midst of the worst winter of the century, with weather and natural gas shortages throwing more than 1.8 million people out of work. One of his first acts was to obtain emergency legislation from Congress to shift gas from state to state. His first major address to the nation in April concerned energy. Public opinion polls showed that at least half of the public did not believe there was an "energy crisis" or believed that it was artificially created. However, the President warned that the United States faced a possible "national catastrophe" unless it responded with the "moral equivalent of war" to dwindling energy supplies by accepting a program based on stringent conservation of fuels, higher energy prices, and penalties for waste.

The detailed National Energy Plan which Carter unveiled to Congress was complex, controversial and, in essence, a massive taxation program. The centerpiece was conservation. To achieve this, his primary proposals were to continue price controls on oil, but to impose a federal production tax to bring prices to world levels, with the government receiving the money and rebating it in some form to consumers; to increase controlled natural gas prices but spread controls to the intrastate market; to force industrial users to convert from oil and gas to coal by taxation; to revise utility rates to save electricity, and to provide incentives to conserve energy in homes. There were almost no provisions or incentives to encourage development of new energy supplies.

Such an unbalanced, inadequate energy plan was bound to face political obstacles. After a year of heated debate, Congress seemed ready to pass watered-down versions of the principal administration proposals with the exception of the federal tax on crude oil, for which there was little Congressional appetite in an election year. By this time, the United States was importing almost half of its oil needs and was running a huge national trade deficit. It imported $45 billion worth of oil in 1977, as compared to $7.7 billion in 1973. The most alarming fact was that public opinion polls showed that half of all Americans didn't know that the nation imported *any* oil. Nor was the nation as a whole aware that importing oil means exporting American jobs.

There was no argument from any source in government or

industry concerning the need to reduce oil and gas consumption through conservation. President Carter called America "the most wasteful nation on earth," for with less than six per cent of the world's population it consumes one-third of the world's energy. What he neglected to add was that America produces one-third of the world's goods and has the highest standard of living. Since the Arab embargo, industry has demonstrated that it can effect 10–15 per cent energy savings through improved efficiency. Homeowners are becoming increasingly conscious of the need to save energy. Nevertheless, conservation does not add to energy supply, which is the key to maintaining the nation's productivity.

The basic fallacy of the Carter program was the assumption that the nation was running out of oil and gas. This is far from the truth. Those who know the most about the nation's oil and gas potential—earth scientists in both government and industry—are in agreement that the United States still has great resources yet to be found onshore and offshore. The United States Geological Survey, which traditionally makes the most conservative estimates, concludes that of our total potential oil resources, *recoverable with present technology*, the United States has exhausted 39 per cent, has a reserve of 24 per cent, and has 37 per cent—or another 100 billion barrels—left to discover if industry is given sufficient incentives for exploration.

Furthermore, the Survey points out that another 150 to 235 billion barrels can be recovered from oil known to be in place by the use of expensive new technology called enhanced recovery. The average recovery of reserves in U.S. oil fields is only about 33 per cent. Improved petroleum reservoir engineering knowledge is increasing recovery by such conventional methods as water flooding and gas injection, but new, presently high cost methods of steam stimulation or flooding and the use of chemicals have enormous potential in recovering known but heretofore unrecoverable oil.

The Carter energy planners did not consult the U.S. Geological Survey when they were preparing their crash program. Instead, they preferred to release a CIA report which said that, in the absence of greatly increased energy conservation, world oil demand would exceed supply by 1985 and the Soviets would be competing with the United States to buy world oil.

As in the case of oil, there is no shortage of potential exploration and producing areas for natural gas. The Potential Gas Committee, under the sponsorship of the Colorado School of Mines, is the most widely accepted non-governmental study drawing on information from government, industry, and academic institutions. In a 1977 study, the Committee estimated that more than 973 trillion cubic feet of probable, possible, and speculative potential reserves remain to be found in the United States. This is 4.5 times current proved reserves, or equal to about 50 times 1976 production.

Furthermore, America has other huge energy resources. There are enough known coal deposits, recoverable under current technological and economic conditions, to last for several hundred years. They could supply at least twice the energy in all the Middle East oil reserves. Also, technology exists to produce oil and gas from coal. There are billions of barrels of oil which can be extracted from vast oil shale deposits in the Western states. America has an estimated six billion pounds of uranium for nuclear power and the great untapped potential of solar energy.

What then is America's *real* energy problem? The development of all these potential resources is mired in a bog of political price controls, environmental restrictions, lack of realistic government priorities and timetable for their use, and the need to recognize the huge amounts of private capital investment necessary to produce new energy supplies.

The concensus of the multitude of government and industry studies is that the United States will be dependent upon oil and gas—whether domestic or imported—for the majority of its energy needs until the turn of the century. Due to technological, economic, and environmental problems, alternate sources of energy will develop slowly. Consequently, the nation must accelerate the development of its own great oil and gas potential if it is to stop the disastrous increase of imports, coming primarily from the Middle East with its ever-present possibility of further wars and Arab oil embargoes.

The nation's future, undiscovered oil and gas fields will be more difficult and costlier to find than in the past. The majority of the country's relatively shallow oil and inexpensive gas has been found. New targets onshore are deeper and increasingly expensive, with an average 5,000-foot wildcat costing $200,000 and many over $1 million. Furthermore, although billions of barrels remain to be discovered onshore, the most prospective new frontiers—with the possibility of giant oil fields—lie offshore California, the Atlantic Coast, in the Gulf of Mexico, and the Gulf of Alaska. The cost of an offshore well starts at $1,500,000 and can be as much as $35 million. Despite great technological improvements in exploration techniques, there is still no direct oil finding method. The odds are even higher today in the search for big oil fields. Since risk capital must come from profits, the price of oil governs the amount of exploration carried out. The Federal Energy Administration estimated in 1976—before the Carter energy plan—that if oil and gas prices were regulated at low levels, exploration would be depressed to the point where imports would rise from over 7 million barrels a day to 13.5 million barrels a day in 1985. However, deregulation of oil and gas prices would increase exploration to the point where imports could be reduced to 5.9 million barrels daily. A higher cost for domestic supplies would not significantly raise the cost of products to the consumer any more than increasing dependence on costly foreign oil and gas.

It has always been a rule of thumb that the oil and gas the nation uses on any given day, someone had to start looking for five years before. Today, environmental restrictions and objectives have lengthened that lead time phenomenally. A prime example is Exxon's giant Hondo field in the Santa Barbara Channel. It was discovered in 1969, a few months following the Santa Barbara oil slick, in a part of the Channel where limited exploration drilling was still permitted. It opened up a new oil area estimated to contain between 1 and 3 billion barrels. In attempting to develop the Hondo field, Exxon has been subjected to twenty major public hearings, preparation of three Environmental Impact Statements, forty-four studies by consultants, and nine lawsuits. In 1978, the field still faced governmental problems and delays with offshore production facilities before production could begin.

Limited exploration offshore California did not resume until four years after the Santa Barbara blowout. Meanwhile, the federal government cautiously resumed lease sales in the Gulf of Mexico and the Gulf of Alaska. Lease sales offshore Atlantic in the Baltimore Canyon trough were not authorized until 1976 and were immediately tied up in a two-year environmental lawsuit. Exploration drilling began in 1978, but even if commercial discoveries are made, required government permits to satisfy environmental requirements will delay production for six to eight years.

Environmentalists, in government and out, are actively pressing legislation which would lock up millions of acres of potential oil and gas land onshore. The Carter administration has introduced a bill to Congress which would set aside 91.7 million acres in Alaska—about one-fourth of the state—as wilderness areas. This means that man can enter, but when he leaves "must leave no trace that he has been there." This would ban all roads, airstrips, lodges, motor boats, etc., and any commercial activity. The areas would be preserved for wildlife and available only to hardy backpackers. Most of these areas are rich in minerals, timber, and commercial resources. Some 9 million acres are on the North Slope, where such withholding would adversely affect further development of oil and gas. Also, the location of these wilderness areas would effectively lock up a major part of the rest of the state for development, since no access roads or pipelines could be built through them to non-wilderness areas. Governor Jay Hammond of Alaska, an environmentalist, and the Alaskan congressional delegation are strongly opposed. They advocate setting aside only 25 million acres under federal protection, with 57 million under federal-state jurisdiction. This plan would permit mineral, energy, and commercial development in the jointly-controlled areas.

The greatest environmentalist threat to oil and gas development onshore in the "lower 48" concerns the most spectacular new oil province, the Overthrust Belt, a major geological feature covering vast, sparsely explored areas in Montana, Idaho, Wyoming, and Utah. After the best efforts of explorationists failed in this area for decades, a group of Ft. Worth independents, American Quasar Petroleum Corp., discovered a giant field in the Utah portion in 1975 and set off one of the biggest booms since the early 1950's. The Overthrust Belt is conservatively estimated to contain some 2 to 3 billion barrels of oil and up to 20 trillion cubic feet of gas. The largest concentration of good acreage for exploration is in national forests, and 47 per cent of the federal lands are already closed to petroleum exploration. Now, the U.S. Forest Service plans to double to 30 million acres the federal lands set aside for wilderness areas, with most of the withdrawals coming from the Overthrust Belt. In addition, the Sierra Club is legally seeking to suspend all oil and gas leases in the area issued since 1970.

Oil and gas industry exploration and drilling technology has increased dramatically since the 1950's in order to meet the difficult new challenges posed by nature both onshore and offshore. Dynamic, innovative concepts and approaches have evolved. However, just as the Lilliputians managed to immobilize Gulliver by tying him up with myriad tiny strings, so have the multitude of haphazard, uncoordinated, conflicting laws, regulations, and legal actions managed to place a giant nation in an astonishing predicament.

The one thing that has not changed is the spirit, ingenuity, perseverance, and vision of the wildcatters who continue to search, no matter what the obstacles placed in their way by nature or men. Today's breed still heed the siren call of Kipling's explorer, hearing "one everlasting Whisper day and night repeated—Something hidden. Go and find it. Go and look behind the Ranges. Something lost behind the Ranges. Lost and waiting for you. Go!"

Such a one is Robert A. Hefner III of Oklahoma City, who, in the tradition of the great wildcatters and scientists, has pioneered a new frontier for the discovery of huge deep gas resources. His roots lie in the Spindletop boom, where his grandfather, fresh out of law school, went to start an oil law firm. He was a lawyer-oilman in early Oklahoma oil boom times and eventually an Oklahoma Supreme Court Justice. His son, Robert A. Hefner, Jr., became an independent oilman. But it was more than his heritage that made Robert A. Hefner III become an explorer.

As a petroleum geology student at the University of Oklahoma in the late 1950's, he studied the stratigraphy of the Anadarko basin, the vast L-shaped structure extending from the Texas Panhandle across the Oklahoma Panhandle and into Kansas which contains one of the world's greatest gas fields. This was the field whose first discovery well in 1918 was located by Charles Gould, founder of the University of Oklahoma's geology department, who pioneered the acceptance of geology as a science and was the first to realize and point out that anticlines are favorable places to look for oil and gas.

As Bob Hefner studied the basin he saw that it was one of the largest continental basins in the world in which the accumulation of potentially productive sediments was completely undeveloped. On the shallow rims of the basin some of the world's largest proven gas reserves existed, and he could not understand, as a fledgling scientist, why the remaining 24,000 cubic miles of undeveloped sediments would not have enormous potential for vast supplies of natural gas. This was contrary to accepted geological thinking, which held that at great depths the weight of the overburden would crush the reservoirs or diminish them to the extent that there would not be sufficient reservoir quality remaining to provide storage space and flow channels for great gas supplies. If there *was* gas at such depths —15,000 to 50,000 feet—there was no technology yet to drill it. Furthermore, the controlled price of gas was so low that, even if technology were developed, gas at such depths could not possibly be commercial because of the great costs involved. Besides, the psychology of the oil industry then was almost entirely oil-related. Nobody was really looking for gas.

None of these obstacles intimidated Bob Hefner. He had

caught his dream. "As an idealist college boy, facing this vast new environment, I thought if it was there, we could find it and bring it into the economic system," he says.

After graduation he worked for Phillips Petroleum, learning the art of seismograph and participating in worldwide geological studies. He went to work for the family oil company, but quit within six months to release the creative drive within that was calling him to the Anadarko basin. With two partners he formed the Glover Hefner Kennedy Company, or GHK, and in 1959 began geological studies. By 1962 the company was acquiring leases and doing seismic testing, but it was five long years before they were able to raise sufficient funds to drill their first deep wildcat. Two deep tests had already been drilled in the basin. Both were dry. So was Bob's.

Bob Hefner knew from studying the efforts of the great wildcatters, whose achievements had been part of his inspiration, that the first dry hole doesn't count. In 1967, he commenced drilling a well, the No. 1 Green, which he was determined to take to depths never before penetrated in the basin. Then began a saga of technological and financing problems. He was ridiculed by many persons, and there were times when he was tempted to give up. But like Dad Joiner with his Daisy Bradford well in East Texas, Hefner always managed to find others who had faith in backing him.

In 1969, after two years of continuous operations, overcoming immense drilling problems and making many record technological breakthroughs, the No. 1 Green reached a total depth of 24,454 feet, then the second-deepest well ever drilled in the world, and found a huge gas reservoir. When it was successfully completed as a producer it was the second-deepest producing reservoir in the Anadarko basin. It was the highest pressure gas well in the world, requiring the largest wellhead that had ever been constructed. In addition, the Dad Joiner of the Anadarko basin had discovered one of the most prolific producing gas wells in the country. Then, in its initial testing, the well blew up, and afterward its production was never the same.

The No. 1 Green was a record breaker and opened up a new frontier. But it was an economic disaster. The well's total cost was $6,500,000—$2 million for leases and seismic information and $4.5 million to drill and complete. The Green gas is sold in regulated interstate commerce to the Chicago area at 22 cents a thousand cubic feet—equal to about $1.50 per barrel of oil and contrasted to $4.50 per thousand cubic feet for imported natural gas. After seven years of production, or the energy equivalent of around 1.5 million barrels of oil, Hefner's company had recouped less than a fourth of its actual investment.

In 1970, after Glover died, Hefner and his partner David O'D. Kennedy took a great gamble on another deep well, the Farrar. Reaching 22,408 feet, the Farrar is the deepest well ever drilled and owned 100 per cent by two individuals. Once again, the geology was right, and large gas reserves were found. But technological problems and low gas prices spelled another economic failure.

The No. 1 Green and Farrar typified what was wrong with natural gas price controls and why there is little incentive to develop resources for an interstate market. Hefner's discoveries encouraged new deep drilling in the basin for the intrastate free market. As a result, the basin's potential deep gas reserves are now estimated to be the energy equivalent of the discovery of six giant oil fields. Their full development for the national market will depend on government pricing policies—either deregulation of natural gas price controls or adequate pricing to justify the risks and high costs involved.

Bob Hefner has continued his exploration in the basin, for

he is optimistically betting on the future. In 1978, he made another technological breakthrough which greatly increases the prospects of successful ultradeep production. A major problem in such deep gas wells is that the gas pressures are so extraordinarily high that efforts to open up the producing formations with conventional fracturing methods and materials have too often failed. Exxon's research laboratory developed a new fracturing fluid, containing bauxite, which Hefner has proved can resist the pressures and open formations for a free gas flow. This has opened up a new frontier within the new frontier. In 1978, more than 200 wells will penetrate the deep formations of the Anadarko basin in search of natural gas. Given proper price incentives, Hefner says some 5,000 deep gas wells could be drilled in the basin in the next decade.

How many more deep Anadarko basins are there? How many more Overthrust Belts? How many as yet unknown, unsuspected accumulations of oil and gas resources remain to be found? The past and the present indicate the answer—as many as there are men who have the incentive to "go and look behind the Ranges" and who seek to find new ways to unravel Nature's secrets. Bob Hefner puts it this way, "To the extent mankind is eternal, resources are infinite."

In 1976 the great geologist and oil finder Wallace Pratt celebrated his ninety-first birthday at his retirement ranch at the foot of the Santa Catalina Mountains near Tucson, from which he carries on a vigorous worldwide correspondence with two generations of petroleum geologists who have succeeded him in the search for oil. In an interview he was asked where America needs to go to find new resources of petroleum. "In 1940, composing the text of my book, *Oil in the Earth*, I recorded the conviction that invariably new oil fields were first discovered in 'the minds of men,'" he said. "Probably I should now update that assertion to 'the minds of men and women.' Once a formidable mental hurdle is negotiated, our own Atlantic continental shelf seems to me to be as promising an unexplored hunting ground as we have ever had the good fortune to encounter. Vast areas are extremely promising off the coasts of Alaska and California. And the exploration of our share of the Gulf of Mexico is far from complete."

Pratt concurred in the thinking of his late colleague, A. I. Levorsen, who, in the 1950's at the peak of America's oil finding, said, "If I were a barrel of oil, comfortably located in a pool, hidden in a trap deep in the ground, the region that would be safest for me—where I might live out another 50 or 100 million years in peace and dignity—would be some country where minerals and exploration are nationalized. The reason is that in such countries there is but one hunter, and the chances of eluding him are far better than are the chances of being discovered. The most dangerous place for me to live, as a barrel of oil, would be in the United States where there are thousands of hunters and each has a different weapon."

In the 1970's the "minds of men and women" and the "thousands of hunters" are still at work. But the American eagle seems to have folded its wings. Still, what was true of the past can be true of the future. Oil and gas are not so much the prisoner of the earth's rocks as of the laws men make. If the eagle has the strength to break the bureaucratic bonds that bind his pinions, it can fly again.

About the Author

THE GRANDDAUGHTER and daughter of independent oilmen, Ruth Sheldon Knowles is a seasoned oilwoman in her own right. She is an internationally known petroleum specialist, writer, foreign correspondent, and lecturer. She has made ten trips around the world studying oil producing and exploration operations in the world's major oil centers.

Mrs. Knowles has served as a petroleum consultant to the United States, Mexican, Venezuelan, and Indonesian governments. In 1941 she was appointed by Secretary of Interior Harold L. Ickes to be a petroleum specialist on his staff. He sent her to South America to make the first U.S. government survey of oil fields and refineries. She was a special consultant to the Venezuelan government on its petroleum law. She spent much time in Cuba on oil exploration projects.

In addition to her books, Mrs. Knowles is a frequent contributor to leading popular magazines, newspapers, and professional journals. She has written radio programs for "Voice of America," a series of radio programs concerning the history of Middle East oil which have been broadcast in Arabic in Middle East oil countries, and has written, produced, and narrated a documentary film on Pertamina, Indonesia's national oil company.

The American Women of Radio and Television selected Mrs. Knowles as "Woman of the Year" in 1961, and in 1962 she was selected as Oklahoma's outstanding woman journalist by Theta Sigma Phi, the national professional fraternity for women in journalism.

Mrs. Knowles, who resides in New York City, is the mother of four children.

Acknowledgments and Recommended Readings

IN THE DETAILED research for this book, in addition to interviewing the majority of the wildcatters and scientists, I consulted hundreds of books, trade magazines, newspapers, and scientific journals. Most of the material about oil, unfortunately, is technical. However, anyone interested in knowing more details about the industry's personalities and industry history would doubtless find these nontechnical books helpful and interesting—they were to me: Allan Nevins, *Study In Power* (2 vols., biography of John D. Rockefeller), Scribner, 1953; Sam Mallison, *The Great Wildcatter* (biography of Mike Benedum), Education Foundation of West Virginia, 1953; John Joseph Mathews, *Life and Death of an Oilman* (biography of E. W. Marland), University of Oklahoma Press, 1952; James A. Clark and Michel T. Halbouty, *Spindletop*, Random House, 1952; Samuel W. Tait, Jr., *The Wildcatters*, Princeton University Press, 1946; Carl Coke Rister, *Oil! Titan of the Southwest*, University of Oklahoma Press, 1949; Ralph W. Hidy and Muriel E. Hidy, *Pioneering in Big Business* (vol. 1, history of Standard Oil Company of New Jersey, 1882–1911), Harper, 1955; George Sweet Gibb and Evelyn H. Knowlton, *The Resurgent Years* (vol. 2, history of Standard Oil Company of New Jersey, 1911–1927), Harper, 1956; Kendall Beaton, *Enterprise in Oil* (history of Shell in the United States), Appleton, 1957; Paul H. Giddens, *Standard Oil Company* (Indiana), Appleton, 1955; Earl M. Welty and Frank J. Taylor, *The Black Bonanza* (history of Union Oil Company of California), McGraw-Hill, 1958; Wallace E. Pratt, *Oil in the Earth*, University of Kansas, 1943.

For a detailed picture of the tumultuous national and international events which have shaped our energy history from 1970 to the present, I recommend my book *America's Oil Famine: How It Happened and When It Will End*, Coward McCann & Geoghegan, 1975.

One of the most rewarding experiences in writing a book is the enthusiastic help which everyone, especially librarians, accords the writer. For *The Greatest Gamblers* I was particularly indebted to the Tulsa Public Library's splendid technical department, headed by Sam Smoot; to the efforts of Eugenia Maddox, in charge of the Library of the University of Tulsa; to the New York Public Library, especially to Robert A. Hug; to Virginia Smythe and Eileen Casey of the American Petroleum Institute's library; and to the Department of the Interior's library (Washington, D.C.) The last contains the most complete collection of oil and gas books in the world, collected by E. B. Swanson, former Petroleum Economist for the Department and a bibliophile who repeatedly called my attention to rare and valuable material.

I want to express my gratitude to my many friends in the industry who so generously gave their time and help in providing information. At critical times in the manuscript's preparation I was guided and helped by Dr. Ben Henneke, former president of the University of Tulsa, and by Elaine St. Johns and Connie Moon.

RUTH SHELDON KNOWLES

New York City

Index